National Conference on Theory of Magnetism 2016

Journal of Physics: Conference Series Volume 827

Hefei, China
8 – 11 March 2016

ISBN: 978-1-5108-4075-1
ISSN: 1742-6588

Printed from e-media with permission by:

Curran Associates, Inc.
57 Morehouse Lane
Red Hook, NY 12571

Some format issues inherent in the e-media version may also appear in this print version.

Printed by Curran Associates, Inc. (2017)

For permission requests, please contact the Institute of Physics
at the address below.

Institute of Physics
Dirac House, Temple Back
Bristol BS1 6BE UK

Phone: 44 1 17 929 7481
Fax: 44 1 17 920 0979

techtracking@iop.org

Additional copies of this publication are available from:

Curran Associates, Inc.
57 Morehouse Lane
Red Hook, NY 12571 USA
Phone: 845-758-0400
Fax: 845-758-2633
Email: curran@proceedings.com
Web: www.proceedings.com

Table of contents

Volume 827

National Conference on Theory of Magnetism

8–11 March 2016, Hefei, China

Accepted papers received: 14 March 2017
Published online: 17 April 2017

Preface

011001
OPEN ACCESS
National Conference on Theory of Magnetism

011002
OPEN ACCESS
Peer review statement

Papers

Invited Paper

012001
OPEN ACCESS
The nature of three dimensions: Non-local behavior in the three-dimensional (3D) Ising model

Z.D. Zhang.....1

Invited Paper

012002
OPEN ACCESS
Novel nonlinear excitations in ferromagnet excited by all-magnonic spin-transfer torque

Zai-Dong Li, Qiu-Yan Li and W. M. Liu.....11

Invited Paper

012003
OPEN ACCESS
Magnetic crossover realized by electrical methods

Hua Hao, XiaoHong Zheng, Ting Jia and Zhi Zeng.....17

Invited Paper

012004
OPEN ACCESS
Micromagnetic simulation of two-body magnetic nanoparticles

Fei Li, Jincheng Lu, Yu Yang, Xiaofeng Lu, Rujun Tang and Z. Z. Sun.....23

Invited Paper

012005
OPEN ACCESS
The collinear ↑↑↓↓ magnetism driven ferroelectricity in double-perovskite multiferroics

Ting Jia, Zhi Zeng and H. Q. Lin.....29

Invited Paper

012006
OPEN ACCESS
Magnetic phase diagram of FeO at high pressure

Peng Zhang, R. E. Cohen and K. Haule.....35

012007
OPEN ACCESS
Distance effect on magnetic coupling of Cr doped Bi_2Se_3: A first-principle and Monte Carlo study

Deyao Wu, Jingyao Liu, Ruqian Lian, Guigui Xu, Kehua Zhong, Jian-Min Zhang and Zhigao Huang.....39

Invited Paper

012008
OPEN ACCESS
The local magnetic moment and electron transfer of ZnO-based dilute magnetic semiconductors

Bin Zhuang, Yanmin Yang, Kehua Zhong, Jian-Min Zhang, Guigui Xu and Zhigao Huang.....44

012009
OPEN ACCESS
Phase diagrams of the Kane-Mele-Hubbard model in the presence of an external magnetic field

Ya-Min Quan, Wei-Cheng Bao, Wei-Hua Wang and Da-Yong Liu.....49

Invited Paper

012010
OPEN ACCESS
Possible three-dimensional superconducting properties of strongly correlated honeycomb compound $In_3Cu_2VO_9$

Shi-qing Jia, Qing-wei Wang, Jing Liu and Liang-Jian Zou.....53

012011
OPEN ACCESS
Spin and valley transport in the ferromagnetic MoS_2 junctions subjected by the gate voltage

P. Ye, R.Y. Yuan, Y. Y. Xia and X. Zhao.....59

Invited Paper

012012
OPEN ACCESS
Spin-polarized electron transport in hybrid graphene-BN nanoribbons

Song Gao, Wei Lu, Guo-Hui Zheng, Yalei Jia and San-Huang Ke.....63

Invited Paper

012013
OPEN ACCESS
Pressure-induced metallic ferromagnetism in antiferromagnetic insulator LaMnAsO

HuaiBao Tang, Guang Li, Qun Yang, XueQing Zuo, HaiJun Zhang and DaYong Liu.....69

012014
OPEN ACCESS
Magnetization and specific heat studies of the low temperature anomalies in terbium iron garnet $Tb_3Fe_5O_{12}$

M Lahoubi, W Wang and V Varazashvili.....75

012015
OPEN ACCESS
The structural properties of $LaRO_3$ (R=Cr, Mn, Fe): a first-principles calculation

Y. R. Li, Z. T. Hou, T. X. Wang, Y. Li, H. Y. Liu, X. F. Dai and G. D. Liu.....79

012016
OPEN ACCESS
Magnetism induced by Ga vacancy in rare earth R(R=Gd,Eu,Tm)doped GaN

Z.T. Hou, Y.R. Li, T.X. Wang, Y. Li, H.Y. Liu, X.F. Dai and G.D. Liu.....85

012017
OPEN ACCESS
Effect of magnetic field strength on M_6C carbide precipitation behavior in $W_6Mo_5Cr_4V_3$ high speed steel during tempering

Y Wu, Z W Zhang, H H Li and X Zhao.....91

012018
OPEN ACCESS
Magnetic Properties and Exchange Bias Effect of Cu and Cr doped Sm_2Co_{17} Alloy Film

Dashun Dong, Qingqing Fang, Wenwen Wang and Jinjin Yang.....95

012019
OPEN ACCESS
Microstructure, magnetic and magneto-optical properties of Fe-doped $SmCo_5$ thin films

W W Wang, Q Q Fang, J J Yang, D S Dong and J G Li.....99

012020
OPEN ACCESS
Effect of high magnetic field on recrystallization behavior of cold rolled H70 brass

Y Wu, Q Wang and X Zhao.....105

IOP Conf. Series: Journal of Physics: Conf. Series **827** (2017) 011001 doi:10.1088/1742-6596/827/1/011001

Preface for the Special Collection

Magnetism and magnetic properties are the important branches of condensed matter physics. The research of magnetism is closely related to the national economy, national defense security and people's daily life. The national conference on the theory of magnetism provides an interactive platform for Chinese researchers to demonstrate their latest achievements in the recent investigations on the theoretical magnetism and related magnetic experiments, and to explore and discuss new ideas, new methods and results so as to promote further development in the theory of magnetism in China. Thirteen workshops have been successfully held since 1964. The 14th National Conference on Theory of Magnetism was held in Hefei, the capital of Anhui province, operated by the Institute of Solid State Physics, Chinese Academy of Sciences, on April 8-11, 2016.

The opening ceremony of the Conference was chaired by Prof. Zhi Zeng from the Institute of Solid State Physics, Chinese Academy of Sciences. Prof. Baonian Wan, the vice president of Hefei Institutes of Physical Science, Chinese Academy of Sciences, and Prof. Wuming Liu, the Chairman of the Organizing Committee, addressed in the ceremony. 232 participants from 69 research institutes and universities attended the Conference, including more than 20 famous experts in the field of magnetism. Five top experimentalists in magnetism were invited to present their latest progresses. The talks were focused on the following seven sessions: spintronics & spin transport, quantum magnetism in nanostructures & low-dimensional systems, first-principles calculations and computational simulation of magnetic materials, quantum theory of magnetism, magnetic domain structures and macroscopic quantum phenomena, metallic & alloy magnetic materials, physics of materials under strong magnetic fields, which were very exciting and aroused strong interests of the participants.

This Conference was very successful. It has made a great breakthrough in scale, distinguished in the overall high quality of academic reports, enthusiastic academic discussions, and active poster presentations, together with careful and orderly

IOP Conf. Series: Journal of Physics: Conf. Series **827** (2017) 011001 doi:10.1088/1742-6596/827/1/011001

arrangements of the Program. During the two-day conference, 50 reports were presented, of which 14 were plenary lectures, 26 were invited talks, and 10 were oral talks. Poster presentation was for the first time set up in conference, with a total of 24, from which three were selected for the best poster awards. In addition, the Conference received a total of 86 abstracts. We acknowledge the editorial office and publishing house of the *Acta Physica Sinica* in assisting the publication of the *Abstract Booklet*. We also thank the British *IOP* Publishing Group for the first time agree to publish a special issue on *Journal of Physics: Conference Series* for these invited and contributed papers submitted to this Conference papers.

The Organizing Committee of the conference approved the establishment of the poster session and publishing of conference papers, suggested to incorporat into the future organizing procedures, and proposed to increase the number of the best poster winners so as to attract more young scientists and postgraduates to participate the conference and share their frontier research work. The National Conference on Magnetism Theory was suggested to fix on the first weekend after the *Qingming Festival* and be held every two year. The 15th Conference held in 2018 will be hosted by Fujian Normal University in Fuzhou.

Wuming Liu Chairman of the Organizing Committee

Zhi Zeng Chairman of the Local Organizing Committee

LiangJian Zou Vice Chairman of the Local Organizing Committee

Peer review statement

All papers published in this volume of *Journal of Physics: Conference Series* have been peer reviewed through processes administered by the proceedings Editors. Reviews were conducted by expert referees to the professional and scientific standards expected of a proceedings journal published by IOP Publishing.

IOP Conf. Series: Journal of Physics: Conf. Series **827** (2017) 012001 doi:10.1088/1742-6596/827/1/012001

The nature of three dimensions: Non-local behavior in the three-dimensional (3D) Ising model

Z.D. Zhang
Shenyang National Laboratory for Materials Science, Institute of Metal Research, Chinese Academy of Sciences, 72 Wenhua Road, Shenyang, 110016, P.R. China

zdzhang@imr.ac.cn

Abstract. In this work, we first reveal the nature of three dimensions: the existence of non-local behavior. Taking the three-dimensional (3D) Ising model as an example, we analyze the transfer matrices as well as the partition function to discuss in details the origin of non-local behavior in three dimensions. We then point out that fail to deal with non-local behavior is one of the main disadvantages of several theoretical approaches (such as conventional low-temperature expansions, conventional high-temperature expansions, perturbative renormalization group and Monte Carlo methods, etc), all of which take local environments into account only. This work provides a deep understanding on the natural difference between 3D bulk materials and 2D materials, and shows that how to treat the non-local behavior and the interaction between spins along the third dimension is a key point for solving exactly the problems of the 3D systems with the nearest neighboring interactions. Finally, we will briefly review recent advance in the 3D Ising model and related topics.

1. Introduction

The trend of study on materials science has flocked to nanostructured materials (such as 1D, 2D materials and their composites) since 1980s. However, the study on the physical properties of 3D bulk materials is still important for application in various fields, such as electronic, energy, etc. Furthermore, the understanding of the 3D materials provides a chance to understand well the 3D world we are living. For a full understanding on a physical system, it is important to solve exactly the model as representation of the system. Unfortunately, even for the simplest model, it has been a great challenge for solving exactly the Ising model in three dimensions.

The study on the Ising models has a long history in fields of both physics and mathematics. After the pioneer work of Onsager [1] for the exact solution of two-dimensional (2D) Ising model, there have been many developments for 2D Ising models and 3D Ising models. In the case of 2D Ising models, it is easy to find out the exact solution following the Onsager's approach [1], when no non-trivial topological structures exist. However, almost all the results for 3D Ising models had been obtained by approximation approaches, such as low-temperature expansions, high-temperature expansions, perturbative renormalization group and Monte Carlo methods, etc. The author conjectured that the non-trivial topological structures of the 3D Ising model can be trivialized in a higher dimensional space and the 3D Ising model can be realized as the free statistic model on them with topological/geometrical phases on eigenvectors [2,3].

In this work, on the observation of the formula of the partition functions, we first reveal the non-local behavior of the 3D Ising model. We then point out that fail to deal with non-local behavior is one of the main disadvantages of several theoretical approaches (such as conventional low-temperature expansions, conventional high-temperature expansions, renormalization group and Monte Carlo methods, etc), all of which take local environments into account only. It is clear now that these approximation approaches cannot serve as the standard for judging the putative exact solution of the 3D Ising model. Finally, we will briefly review recent advance in the 3D Ising model and related topics.

2. Non-local behavior of 3D Ising model

As in [2,3], we consider the Ising Hamiltonian on an orthorhombic lattice in 3D Euclidean space, with up-spin or down-spin at each lattice point:

$$H = -J\sum_{\tau=1}^{n}\sum_{\rho=1}^{m}\sum_{\delta=1}^{l} s_{\rho,\delta}^{(\tau)} s_{\rho,\delta}^{(\tau+1)} - J'\sum_{\tau=1}^{n}\sum_{\rho=1}^{m}\sum_{\delta=1}^{l} s_{\rho,\delta}^{(\tau)} s_{\rho+1,\delta}^{(\tau)} - J''\sum_{\tau=1}^{n}\sum_{\rho=1}^{m}\sum_{\delta=1}^{l} s_{\rho,\delta}^{(\tau)} s_{\rho,\delta+1}^{(\tau)}. \quad (1)$$

Here only the nearest neighboring interaction between spins at each lattice point is considered. We follow the notation of Onsager-Kaufman-Zhang for Pauli matrices [1-4]: $s'' = \begin{bmatrix} 0 & -1 \\ 1 & 0 \end{bmatrix}$ $(= i\sigma_2)$, $s' = \begin{bmatrix} 1 & 0 \\ 0 & -1 \end{bmatrix}$ $(= \sigma_3)$, $C = \begin{bmatrix} 0 & 1 \\ 1 & 0 \end{bmatrix}$ $(= \sigma_1)$, and we have the relation $\tanh K^* = e^{-2K}$. We introduce $K = \beta J$, $K' = \beta J'$ and $K'' = \beta J''$ with $\beta = 1/k_B T$. The partition function of the 3D Ising model can be given as follows [3]:

$$Z = (2\sinh 2K)^{\frac{m \cdot n \cdot l}{2}} \cdot trace(V_3 V_2 V_1)^m \equiv (2\sinh 2K)^{\frac{m \cdot n \cdot l}{2}} \cdot \sum_{i=1}^{2^{n \cdot l}} \lambda_i^m \quad (2)$$

$$V_3 = \prod_{j=1}^{nl} \exp\{-iK''\Gamma_{2j}\left[\prod_{k=j+1}^{j+n-1} i\Gamma_{2k-1}\Gamma_{2k}\right]\Gamma_{2j+2n-1}\} = \prod_{j=1}^{nl} \exp(K'' s'_j s'_{j+n}); \quad (3)$$

$$V_2 = \prod_{j=1}^{nl} \exp\{-iK'\Gamma_{2j}\Gamma_{2j+1}\} = \prod_{j=1}^{nl} \exp(K' s'_j s'_{j+1}); \quad (4)$$

$$V_1 = \prod_{j=1}^{nl} \exp\{iK^* \cdot \Gamma_{2j-1}\Gamma_{2j}\} = \prod_{j=1}^{nl} \exp(K^* C_j). \quad (5)$$

Here we introduce the following generators of Clifford algebra of 3D Ising model:

$$\Gamma_{2j-1} = C \otimes C \otimes \ldots\ldots \otimes C \otimes s' \otimes 1 \otimes \ldots \otimes 1 \quad (\text{j times } C) \quad (6a)$$

$$\Gamma_{2j} = C \otimes C \otimes \ldots\ldots \otimes C \otimes (is'') \otimes 1 \otimes \ldots \otimes 1 \quad (\text{j times } C) \quad (6b)$$

The Clifford algebra is generated by the Pauli matrices. Clearly, we have the following formulas:

$$i\Gamma_{2j-1}\Gamma_{2j} = 1 \otimes 1 \otimes \ldots\ldots \otimes 1 \otimes C \otimes 1 \otimes \ldots \otimes 1 \tag{7}$$

$$s'_j s'_{j+1} = 1 \otimes 1 \otimes \ldots \otimes 1 \otimes s' \otimes s' \otimes 1 \otimes \ldots \otimes 1 \tag{8}$$

$$s'_j s'_{j+n} = 1 \otimes 1 \otimes \ldots \otimes 1 \otimes s' \otimes 1 \otimes \ldots \otimes 1 \otimes s' \otimes 1 \otimes \ldots \otimes 1 \tag{9}$$

The problem becomes to how to deal with the trace of the transfer matrices $V = V_3V_2V_1$, specially, V_3 with non-linear terms of Γ matrices. There are direct products of (n-1) terms of unit matrices in between two Pauli matrices s' in Eq. (9).

It is noticed that the transfer matrix V_1 or V_2 with the product of two Γ matrices has trivial topologic structure, which can be represented as Lie group for a rotation. The transfer matrix V_3 is the root of difficulties to hinder the direct application of Onsager-Kaufmann approach to 3D Ising model, due to the high-order terms of Γ-matrices. The non-trivial topological effect as well as the non-local behavior can be seen also in the language of the spin (σ) variables although all expressions seems local. Although the formulas for the transfer matrices V_2 and V_3, i.e., Eqs. (3) and (4), look very similar, they are quite different. The difference in the subscripts of s' in Eqs. (3) and (4) represents different meanings, as shown in Eqs. (8) and (9). $s'_j s'_{j+1}$ represents the interaction between the most neighboring spins along the second dimension in a plane, while $s'_j s'_{j+n}$ represents the interaction between the most neighboring spins along the third dimension. The latter term looks like an interaction between two spins located far from each other, via a chain of n spins in the plane (actually, via nm spins in the plane if one considers the periodic condition already used along the first dimension, as shown in Eq. (2)). The reason is that the third transfer matrix V_3 must follow the sequence of the spin (σ) variables arranged and fixed already in the first two transfer matrices V_1 and V_2. According to this fixed order, although the interaction between a spin and one of its neighboring spins along the third dimension is the nearest neighboring one, its effect is correlated with the states of n other spins in the plane. It is equivalent effectively to a long-range and many-body interaction in which (n+1) spins are involved. The same effect happens for every interaction along the third dimension in the 3D Ising model. Therefore, the non-local correlation and the global effect indeed exist in the 3D Ising system. From another angle of points of view, all the spins in the 3D Ising model are entangled. With the language of the formionic (Γ) variables (see Eq. (3)), it is easy to obtain the same conclusion.

It was understood that the different topological states (e.g. knots/links) formed by up or down spins also contribute to the partition function and hence the free energy and other physical properties of the system [3].Therefore, the partition function of the 3D Ising model can be written in the following two parts as:

$$Z = Z_{Local} \cdot Z_{Non-local} \tag{10}$$

Here Z_{Local} and $Z_{Non-local}$ represent the contributions from the local environment and the non-local behavior to the partition function of the 3D Ising system. Correspondingly, the free energy of the 3D Ising model can be written as the sum of the local and non-local parts:

$$F = F_{Local} + F_{Non-local} \tag{11}$$

Here F_{Local} and $F_{Non-local}$ represent the contributions from the local environment and the non-local behavior to the free energy of the 3D Ising system. Consequently, each of the physical properties, such as specific heat, magnetization, susceptibility, correlation function, etc., can be treated as the sum of local and non-local parts.

3. Disadvantages of several theoretical approaches

Attempts to apply the algebraic method used for solving the 2D model to the 3D problem are seriously hindered at an early stage, because the operators of interest generate a much large Lie algebra being so large that it would seem to be of little value [5]. The combinatorial method developed by Kac and

Ward [6] introduces some problems in topology. This combinatorial method of counting the closed graph cannot be generated in any obvious way to the 3D problem, since the peculiar topological property is that a polygon in three dimensions does not divide the space into an "inside and outside" [5]. These problems are what an approach for the 3D Ising model must face. It is clear that any approaches based on only local environments, such as conventional low-temperature expansions, conventional high-temperature expansions, perturbative renormalization group, Monte Carlo simulations, etc, cannot be exact for the 3D Ising model, even though they may be exact for the 2D cases [2,3]. We will discuss the disadvantages of these theoretical techniques as follows (for details, the readers refer to [2,3]):

3.1 Low-temperature expansions

All systematic methods for the determination of series coefficients are at some level graphical or diagrammatic. With each coefficient is associated a set of graphs of some given topological type. To each graph corresponds a numerical contribution according to a well-defined rule. To calculate the required coefficient, one simply sums all contributions. For conventional low-temperature expansions, the appropriate choice is to define the partition function such that the fully aligned spin-up state is taken as having zero energy, since the low-temperature series is a perturbation expansion about this state [5,7-10]. The conventional low-temperature expansions, evaluated by systematically overturning spins from the ground state with all spins "up", consider only the local environment of spins and their interactions, neglecting the non-local behavior of the 3D Ising model (i.e., the interaction via the chain of n spins in the plane). Therefore, the conventional low-temperature expansions take into account only the local part of the partition function and the free energy of the system. It gives the same fundamental leading term for the 2D triangular Ising lattice and the 3D simple cubic lattice, because it relates directly with the Ising model lattice of coordination number q, regardless of what the dimension it is. This is also the origin of the divergence of the low-temperature expansions for the 3D Ising model. The reason for the zero radius of the circle of convergence of the conventional low – temperature series for the 3D Ising model is the direct result that the non-local effect lacks in the series. It is noticed that the conjectured solution for the 3D Ising model agrees well with C.N. Yang's magnetization for 2D. If one accepted the conjectured exact solution in [2], which consists of both local and non-local parts of the physical properties of the 3D Ising system, one would derive the non-local contribution by extracting the local part from the total property.

3.2 High-temperature expansions

It is known that the conventional high - temperature series expansion is an exact expansion for the 2D Ising model and it is easy to write down every terms of the high - temperature series expansion, by accounting the loops (polygons) in the 2D lattice. The loops are the boundary between the areas of spin up and spin down [5,7,11]. However, for the 3D Ising model, one first meets the challenge to write down every terms of the high - temperature series expansion. A basic difficulty is that in 3D, the boundaries between domains of spin up and spin down are not polygons, but polyhedrons. The polygons cannot separate the domains of spin up and spin down. Furthermore, the existence of the non-trivial topological objects (such as Mobius band, Klein bottle, crosses and knots/links, etc.) in the 3D Ising lattice makes the well – known topologic troubles. Actually, any polygons are local, which cannot take into account the non-local behavior of the 3D Ising system, which involves all the states of nm spins in a plane (for an interaction between the neighboring spins along the third dimension). Because of the duality between the high-temperature and low-temperature expansions, the divergence of the conventional low-temperature expansions corresponds to the zero radius of the convergence of the conventional high – temperature series. It was the observation of the author [2,3] that there exists a topological phase transition at/near infinite temperature for the 3D Ising model. In the 3D Ising system, the trivial topological structure at infinite temperature is changed to be the non-trivial topological structure below infinite. The existence of such a topological phase transition makes the possibilities of the existence of several different high-temperature expansions. Namely, the high-temperature expansions are not unique, and the correct one should be that including the non-local behavior.

3.3 Renormalization group

Although the renormalization group theory is described in mathematical terms, it is not rigorous. Although the renormalization group theory has been widely applied to study the critical phenomena, and although it shows the results with high precision for 2D Ising model (in comparison with the Onsager's exact solution) [12-17], it possesses the disadvantages similar to those of the conventional series expansions. During the renormalization group procedures for 3D Ising model, various approximations, such as expansions, perturbations, linearizations, normalizations, etc. are performed. In all the ways, the disadvantages, similar to those of conventional low – and high – temperature expansions, have not been removed. The real-space renormalization group techniques are concerned with the construction of new models from old by averaging dynamical variables of the old model to form the block variable of the new one. The final results depend sensitively on how to divide Kadanoff blocks, define the effective Hamiltonian, determine the details of the block variables, and calculate approximately the partial trace. The larger the Kadanoff blocks, the more accurate the calculations, but the more complicated the procedure. The final results would approach the exact solution if and only if the size of the Kadanoff blocks were chose to be infinite and the infinite terms of the expansion were remained. This is impossible to be done, because much more variables would emerge with increasing the size of the Kadanoff blocks and taking into account more terms of the expansion, which make the calculations become extremely difficult. A serious trouble is that the non-local behavior in 3D involves in the interaction of long-range-like type and all the possible states of nm spins in a plane ($n \to \infty$ and $m \to \infty$) even for only one interaction between two spins along the third dimension, which make any procedure with finite blocks cut the plane of nm spins with the non-local effect. The field theoretical or k – space renormalization group techniques meets the same difficulty that cannot be overcome, since at the deeper level there is a close connection, because the basic idea of them is the same and because there are connections between ϕ and a set of block variables. It is clear now that the perturbative renormalization group with finite blocks cannot take into account the non-local behavior of the 3D Ising model. Therefore, one has to develop the nonperturbative renormalization group for studying the critical phenomena of the 3D Ising model.

3.4 Monte Carlo methods

It is known that Monte Carlo simulations are powerful techniques for numerical calculations [16,18-21], which numerically evaluate canonical thermal averages of some observable A by an approximate one, where M states $\{x_\mu\}$ are selected by importance – sampling process. Any computer simulations including Monte Carlo method for the 3D Ising model are limited by the size effects since the number of the configurations of the 3D Ising model increases in a fashion of 2^N, with the number of the spins or atoms, $N \to \infty$, Usually, the period boundary condition is used as that the simulation is performed on a small system with finite size. This technique breaks down the long-range many-body interaction between the two nearest neighboring spins along the third dimension in the 3D Ising model, which is due to the non-local behavior. As mentioned above, the interaction between the two spins along the third dimension depends on all the possible configurations 2^{nm} of nm spins in a plane. The number nm of the spins in the plane can be infinite, which makes the simulation on any finite sub-systems plus the periodic boundary condition meaningless. This results in that for the finite system Monte Carlo simulation there is not a sharp transition between zero and non-zero magnetization or a sharp peak in the specific heat. Therefore the critical point cannot be located precisely, which hinders to understand its singularity analytically at the critical point. The limitations of the Monte Carlo simulations were summarized in [2]. For the 2D Ising model, the Monte Carlo simulations give much better results, not only because much larger lattice sizes can be dealt with and but also because there is no non-local problem in 2D. Although the combination of the Monte Carlo simulations and the renormalization group techniques reduces the calculation time and give better results by allowing us to study much larger than those possibly by the direct summation methods, the difficulties above of the Monte Carlo simulations are not removed essentially.

4. Some recent advances in 3D Ising model

In this section, we will review some recent advances in the 3D Ising model and related topic.

4.1 Fractals and chaos in 3D Ising lattices

Inspired by the Zhang's generation of Onsager's approach to the 3D Ising model [2], Lawrynowicz et al. discussed the 3D Ising lattices in the context of possible applications of the Jordan-von Neumann-Wigner approach [22]. They reformulated the algebraic part of the theory in terms of the quaternionic sequence of Jordan algebras and looked at some of the geometrical aspects of the 3D Ising lattices [22,23]. They discussed the relationship with Bethe-type fractals, Kikuchi-type fractals and fractals of the algebraic structure, and also the duality for fractal sets and lattice models on fractal sets. It was found that it is possible to have a simpler description in terms of fractals corresponding to algebraic structure involving the quaternionic sequence of Jordan algebras [22]. It was suggested that the commutatively of the 3D Ising model can be realized by application of Jordan algebras in relation with the Clifford algebras. Moreover, the quaternionic sequence of Jordan algebras has the mathematical basis of the quantum mechanics. In [24-26], Lawrynowicz et al. continued to modify and simplify the Ising-Onsager-Zhang procedure for analyzing the 3D Ising lattices by considering some fractal structures in connection with Jordan and Clifford algebras and by following the Jordan-von Neumann-Wigner approach. They concentrated on duality of complete and perfect Jordan-von Neumann-Wigner systems, particular ternary systems, and analyzed algebras of complete Jordan-von Neumann-Wigner systems. It was proven that in the case of a composition algebra, there is a self-dual perfect Jordan-von Neumann-Wigner system related to quaternion or octonion algebras [24]. It was pointed out [22,23,25] that the Onsager-Zhang approach reformatted with the use of the quaternionic sequence of Jordan algebra, is closely related to Heisenberg's approach to quantum theories. The Jordan structures are closely related to some types of fractals, in particular, fractals of the algebraic structures [25]. The study in [25] included fractal renormalization, the renormalized Dirac operator, meromorphic Schauder basis and hyperfunctions on fractal boundaries.

Scaling properties near the critical point indicates the existence of self-similarity behavior for the critical phenomena. Although the 3D Ising system considered is not a truly dynamic one, Zhang and March proposed a specific set of relations between fractal dimensions and critical exponents in the Ising model of statistical mechanics [27]. In particular, we put forward, corresponding to six critical exponents for the Ising model, six fractal dimensions. Assuming the latter proposals, we then derived relationships between such fractal dimensions.

4.2 Critical exponents in 3D Ising models and related systems

The 3D Ising universality for critical indices in magnets and at fluid-fluid phase transition was reviewed in [28]. Experimental data for critical exponents in magnetic materials were compared with theoretical results on the 3D Ising model, as derived based on two conjectures [2]. It was found that critical exponents in some bulk magnetic materials indeed form a 3D Ising universality. The attention was then focused on the critical indices at fluid-fluid phase transition. It is commonly accepted that pure fluids at gas-liquid critical points, and binary liquid mixtures at liquid-liquid critical points belong to the same universality class as the Ising model. On the other hand, the analogy between magnetic behaviour near criticality and the corresponding liquid – vapour behaviour was used to discuss the 2D – 3D crossover in the latter case [29]. The experimentally observed magnetic behaviour near criticality was considered for the ferromagnet $CrBr_3$ to allow fingerprints of the 3D Ising Hamiltonian to be anticipated. The magnetic equation of state for the 3D Ising model near criticality was proposed for $CrBr_3$ [29,30]. Moreover, the dimensionality and the critical exponents of the models for polymer growth in solution and the dynamic epidemic model were discussed in [31]. We also predicted the critical exponents for the Potts model in three dimensions and the percolation exponents (q = 1) for d = 1, 2, 3, 4, 5, and d ≥ 6.

A relationship between the exponents γ and η at liquid – vapour critical point was proposed via the dimensionality d [32]. Similarities and contrasts between critical point behavior of heavy fluid alkalis and d-dimensional Ising model were discussed in [33]. Crucial combinations of critical exponents were given for liquids-vapour and ferromagnetic second-order phase transitions [34]. A theory of critical exponents was developed in terms of dimensionality d plus universality class n [35]. March also presented a unified theory of critical exponents generated by the Ising Hamiltonian embracing solely the discrete values d = 2, 3 and 4, in terms of the critical exponent η [36]. The formula contains the exact values for d = 2 and 4, while for d = 3, it yields the value obtained eerily by Zhang [2]. Furthermore, statistical - mechanical models with separable many-body interactions, especially partition functions and thermodynamic consequences, for a variety of classical and quantum phase transition, were reviewed in [37,38].

Power law singularities and critical exponents in n-vector models were considered within a theoretical approach called grouping of Feynman diagrams (GFD theory [39]. Predictions for corrections to scaling of the perturbative renormalization group and GFD approaches are different. A nonperturbative proof was provided, supporting corrections to scaling of the GFD theory. Highly accurate experimental data very close to the λ-transition point in liquid helium, as well as the Goldstone mode singularities in n-vector spin models, evaluated from Monte Carlo simulation results, are discussed. The analysis showed that in both cases the data can be well interpreted within GFD theory. The singularity of specific heat C_V of the 3D Ising model was studied based on Monte Carlo data for lattice sizes L ≤ 1536 [40]. A direct estimation from $C_V^{\max}$ data suggests that α/ν has a small value (e.g., $\alpha/\nu = 0.113$). The conventional power-law scaling ansatz can be questioned and the data are well described by certain logarithmic ansatz. Corrections to scaling in the 3D Ising model were studied based on non-perturbative analytical arguments and Monte Carlo simulation data for different lattice sizes L [41].

There is a new development in the 3D Ising model, using convex optimization of c-parameter within the conformal bootstrap approach to the four-point correlation functions [42]. In particular the estimates of $\eta = 0.036$ and $\nu = 0.629$ were reported in [42]. However, it was pointed out in [41] that these estimates in [42] were obtained based on certain hypothesis (e.g. the existence of a sharp kink) and that if these hypothesis are not used, then the conformal bootstrap analysis appears to be consistent with the GFD values $\eta = 1/8$ and $\nu = 2/3$. It is noticed that these GFD values are consistent with the solutions obtained in [2].

One- and two-phase isochoric heat capacities C_V and saturated liquid and vapour densities of sec-butanol near the critical point were measured with a high-temperature and high-pressure nearly constant-volume adiabatic calorimeter [43]. The measured thermodynamic properties of sec-butanol near the critical point were interpreted in terms of the complete scaling theory of critical phenomena and the theory of logarithmic singularity of C_V.

4.3 Temperature-time duality and time

Zhang and March presented a detailed analysis of temperature-time duality in the 3D Ising model, by inspecting the resemblance between the density operator in quantum statistical mechanics and the evolution operator in quantum field theory, with the mapping $\beta = (k_BT)^{-1} \rightarrow it$ [44]. We pointed out that in systems like the 3D Ising model, for the nontrivial topological contributions, the time necessary for the time averaging must be infinite, and being comparable with or even much larger than the time of measurement of the physical quantity of interest. The time averaging is equivalent to the temperature averaging. The phase transitions in the parametric plane (β , it) are discussed, and a singularity (a second-order phase transition) was found to occur at the critical time t_c, corresponding to the critical point β_c (i.e, T_c). It is necessary to use the 4-fold integral form for the partition function for the 3D Ising model. The time is needed to construct the (3+1)D framework for the quaternionic sequence of Jordan algebras, in order to employ the Jordan-von Neumann-Wigner procedure. We then turned to discuss quite briefly temperature-time duality in quantum-chemical many-electron theory. We found that one can use the known one-dimensional differential equation for the Slater sum S(x, β) to write a corresponding form for the diagonal element of the Feynman propagator, again with the mapping $\beta \to$ it.

Based on the quaternionic approach developed in [2] for the three-dimensional (3D) Ising model, we study conformal invariance in three dimensions. The 2D conformal field theory was generalized to be appropriate for three dimensions, within the framework of the quaternionic coordinates with complex weights [45]. The Virasoro algebra still works, but for each complex plane of quaternionic coordinates. The quaternionic geometric phase appears in quaternionic Hilbert space as a result of diagonalization procedure which involves the smoothing of knots/crossings in the 3D many-body interacting spin Ising system. Possibility for application of conformal invariance in three dimensions on studying the behaviour of the world volume of the brane, or the world sheet of the string in 3D or (3+1)D, was discussed.

Zeng discussed the possibility of emergence of time as the holographic dimension of gauge systems in Euclidean space [46], which take statistic, e.g. Ising model as concrete implementations. By identifying the renormalization group flow of statistic models with the time flow of dual gravities, he got a universe whose evolution history is qualitatively the same as our real world (also for 3D Ising model). He also commented highlights projected by this idea on the cosmological constant problem and developed preliminary evidences for the validity of this idea. It is interesting to note that the partition function obtained in [2] for 3D Ising model was used for calculation of the entropy.

5. Summary

In summary, we have investigated the non-local behavior of the 3D Ising model. The interaction between the nearest neighboring spins along the third dimension is effectively equivalent to the long-range many-body interaction in which (n+1) spin are involved. This nonlocal property is an intrinsic property of the 3D Ising model, which could cause that any approximation techniques (including the conventional low-temperature expansions, the conventional high-temperature expansions, the perturbative renormalization group and Monte Carlo method) taking into account only the local property cannot be exact for the 3D Ising model, though these approaches work well for other 2D or 1D models. Any approaches taking into account only the local environments (such as local alignments of spins in low-temperature expansions, polygons in high-temperature expansions, finite blocks in renormalization group, or finite unit cells plus the periodic boundary condition in Monte Carlo method) are inconsistence with the global effect induced by the chain of n spins. The exact solution of the 3D Ising model must be calculated by evaluating the partition function of the whole system involving all the possible states of all the spins of the whole 3D system. Therefore, the systematical errors exist seriously in the approximation techniques, which are caused by their disadvantages. These systematical errors of these approximation techniques are related directly to the physical conceptions/pictures at the first beginning and the neglects of important non-local factors during procedures. The systematical errors are intrinsic, which cannot be removed by the efforts of improving technically the precision. Finally, we reviewed briefly some recent advances in the 3D Ising model.

Acknowledgements

This work has been supported by the National Natural Science Foundation of China under grant number 51331006 and 51590883.

References

[1] Onsager L, 1944 *Phys. Rev.* **65,** 117.

[2] Zhang Z.D, 2007 *Phil. Mag.* **87,** 5309.

[3] Zhang Z.D, 2013 *Chinese Phys.* **B 22**, 030513.

[4] B. Kaufman, 1949 *Phys. Rev.* **76**, 1232.

[5] Newell G.F. and Montroll E.W., 1953 *Rev. Mod. Phys.* **25**, 353..

[6] Kac M. and Ward J.C., 1952 *Phys. Rev.* **88**, 1332.

[7] Domb C., 1974 in Phase Transitions and Critical Phenomena, C. Domb and M.S. Green, eds. Vol. 3, (Academic Press, London).

[8] Guttmann A.J. and Enting I.G., 1993 *J. Phys.* A **26**, 807.

[9] Domb C. and Guttmann A.J, 1970 *J. Phys.* **C 3**, 1652

[10] Bhanot G., Creutz M., Horvath I., Lacki J.and Weckel J., 1994 *Phys. Rev.* **E 49**, 2445.

[11] Fisher M.E., 1967 *Rep. Prog. Phys.*, **30**, 615.

[12] Wilson K.G., 1971 *Phys. Rev.* **B 4**, 3174.

[13] Wilson K.G. and Kogut J., 1974 *Phys. Rep.* **12**, 75.

[14] Kadanoff L.P, Götze W., Hamblen D., Hecht R., Lewis E.A.S., Palciauskas V.V., Rayl M., Swift J., Aspnes D. and Kane J., 1967 *Rev. Mod. Phys.,* **39**, 395.

[15] Kadanof L.P., 1975 *Phys. Rev. Lett.,* **16**, 1005.

[16] Binney J.J., Dowrick N.J., Fisher A.J. and Newman M.E.J., 1992 The Theory of Critical Phenomena, An Introduction to the Renormalization Group, (Clarendon Press, Oxford).

[17] Fisher M.E., 1998 *Rev. Mod. Phys.,* **70**, 653.

[18] Binder K. and Luijten E., 2001 *Phys. Rep.,* **344**, 179.

[19] Pelissetto A. and Vicari E., 2002 *Phys. Rep.* **368**, 549.

[20] Müller-Krumnhaar H. andBinder K., 1973 *J. Stat. Phys.,* **8**, 1.

[21] Stoll E., Binder K. and Schneider T., 1973 *Phys. Rev.* **B 8**, 3266

[22] Ławrynowicz J., Marchiafava S. and Nowak-Kępczyk M., 2008 Trends in Differential Geometry, Complex analysis and Mathemtical Physics, Proceedings of the 9th International Workshop on Complex Structures, Integrability and Vector Fields, Sofia, Bulgaria, 25 - 29 August 2008 (pp 156-166). edited by K. Sekigawa, V. S. Gerdjikov and S. Dimiev (World Scientific, Singapore), DOI 10.1142/9789814277723_0018.

[23] Ławrynowicz J., Marchiafava S. and Niemczynowicz A., 2010 *Adv. Appl. Clifford Alg.* **20**, 733.

[24] Lawrynowicz J., Suzuki O. and Niemczynowicz A., 2012 *Adv. Appl. Clifford Alg.* **22,** 757.

[25] Lawrynowicz J., Nowak-Kepczyk M., Suzuki O., 2012 *Inter. J. Bifurcation Chaos,* **22**, **1230003**.

[26] Lawrynowicz J., Suzuki O. and Niemczynowicz A., 2013, *Inter. J. Nonlinear Sci. Numer. Simul.* **14,** 211.

[27] Zhang Z.D. and March N.H., 2012 *J. Math. Chem.* **50**, 920.

[28] Zhang Z.D. and March N.H., 2011 *Phase Transitions*, **84**, 299.

[29] March N.H. and Zhang Z.D.. 2009 *Phys. Chem. Liquids.* **47**, 693.

[30] March N.H. and Zhang Z.D.. 2009 *Phys. Lett.* **A 373**, 2075.

[31] Zhang Z.D. and March N.H., 2010 *Phys. Chem. Liquids,* **48**,403.

[32] Zhang Z.D. and March N.H., 2011 *Phys. Chem. Liquids,* 49, 270.

[33] March N.H.,2014 *Phys. Lett.* **A 378**, 254.

[34] March N.H. 2014 *Phys. Chem. Liquids,* **52**, 697.

[35] March N.H., 2015 *Phys. Lett.* **A 379,** 820.

[36] March N.H. 2016 *Phys. Chem. Liquids,* **54**, 127.

[37] March N.H. and Zhang Z.D, 2010 *J. Math. Chem.* **47**, 520.

[38] March N.H. and Zhang Z.D, 2013 *J. Math. Chem.* 51, 1694.

[39] Kaupuzs J, 2012 *Canadian J. Phys.* **90**, 373.

[40] Kaupuzs J, Melnik R.V.N, Rimsans J, 2013 *Commun. Computat. Phys.* **14**, 355.

[41] Kaupuzs J, Melnik R.V.N, Rimsans J, 2014, arXiv:1407.3095

[42] El-Showk S, Paulos E.F, Poland D, Rychkov S, Simmons-Duffin D, Vichi A, 2014 *J. Stat. Phys.,* 157, 869.

[43] Radzhabova, L. M.; Stepanov, G. V.; Abdulagatov, I. M.; Shakhbanov, K. A, 2013, *Phys. Chem. Liquids,* **51**, 75.

[44] Z.D. Zhang and N.H. March, 2011 *J. Math. Chem.,* **49,** 1283

[45] Z.D. Zhang and N.H. March, 2012, *Bull. Soc. Sci. Letters Łódź Sér. Rech. Déform.* 62: 35. arXiv:1110.5527.

[46] Zeng D.F, arXiv: 1407.8504.

Novel nonlinear excitations in ferromagnet excited by all-magnonic spin-transfer torque

Zai-Dong Li[1], Qiu-Yan Li[1], and W. M. Liu[2]

Abstract. We report the novel nonlinear excitations, such as Akhmediev breathers solution and Kuznetsov-Ma soliton caused by a spin wave passing through a magnetic soliton. The former case demonstrates a spatial periodic process of a magnetic soliton. The other case shows a localized process of the spin-wave background. In the limit case, we get the novel rogue waves with high magnon density distribution and clarify its formation mechanism.

[1]Department of Applied Physics, Hebei University of Technology, Tianjin 300401, China
[2]Beijing National Laboratory for Condensed Matter Physics, Institute of Physics, Chinese Academy of Sciences, Beijing 100080, China

E-mail: lizd@hebut.edu.cn

1. Introduction

Nonlinear excitations [1, 2] are general phenomena in magnetic ordered materials. During the past several decades there has been significant progress in describing the dynamics of magnetic excitations in magnetic nanostructures, such as spin waves, domain wall, soliton and magnetic vortex. The deviation of magnetization from the ground state results in the excitation of spin waves in ferromagnet. Their attractive interaction and instabilities contribute to the existence of topological and dynamic solitons. A magnetic domain wall forms a spatially localized configuration of magnetization in ferromagnet, and it can be seen as a potential hill which separates two generated magnetic states. Therefore, The dynamics of domain wall is of great significance in ferromagnetic nanowires for its potentially technological applications [3, 4, 5, 6, 7, 8]. On the other hand, the dynamic soliton describes the localized states of magnetization which can be reduced to a uniform magnetization by continuous deformation. So, this excited ferromagnet makes a transition to the ground state and it is sometimes said to be topologically equivalent to the ground state. In general, the external field [9] and a spin-polarized current [10] are the main driven forces for the dynamics of magnetization in ferromagnet. The Walker solution analysis [11, 12, 13] has been extensively adopted to investigate the moving domain wall in response to a magnetic field [14] or spin-polarized current [15, 16, 17]. Nowadays, spin polarized currents are commonly used to create, manipulate, and control nanoscale magnetic excitations such as domain walls [18, 19, 20] and vortices [21, 22, 23].

However, the nonlinear excitations have not been well explored. When a spin wave passes through a magnetic soliton, a spin angular momentum can be transferred from the propagating magnons to the soliton which is called by all-magnonic spin-transfer torque [24]. This all-magnonic spin-transfer torque can affect the dynamics of magnetization and novel magnetic states can occur. In this paper, we report the breather solutions and the novel magnetic states. The dynamics of magnetization can be described by the dimensionless Landau-Lifshitz equation

[15, 16] with the spin-transfer torque

$$\frac{\partial \mathbf{m}}{\partial t} = -\mathbf{m} \times \mathbf{h}_{eff} + A_J \frac{\partial \mathbf{m}}{\partial x}, \tag{1}$$

where $\mathbf{m}$ is the normalized magnetization, and the effective magnetic field $\mathbf{h}_{eff}$ takes the form $\mathbf{h}_{eff} = \partial^2 \mathbf{m}/\partial x^2 + [(h_K - 4\pi)\, m_z + h_{ext}]\mathbf{e}_z$. The last term denotes the adiabatic spin-transfer torque, $A_J = b_J t_0/l_0$ with the characteristic time t_0 and length l_0, where $b_J = P j_e \mu_B/(eM_s)$, P is the spin polarization of the current, j_e is the electric current density and flows along the x direction, μ_B is the Bohr magneton, e is the magnitude of electron charge.

2. Exact breather solutions and rogue waves

In order to obtain the exact breather solutions and rogue waves we only consider the isotropic case, i.e., $\partial \mathbf{m}/\partial t = -\mathbf{m} \times \partial^2 \mathbf{m}/\partial x^2$. In this case, the spin wave takes the form $\mathbf{m}_0 \equiv (m_{01}, m_{02}, m_{03}) = (\cos\delta, \sin\delta, 0)$ with $\delta = k_s x$. With the developed Darboux transformation we obtain the exact solutions [25]

$$\mathbf{m} \cdot \sigma = K\,(\mathbf{m}_0 \cdot \sigma)\, K^{-1}, \tag{2}$$

where σ is pauli matrix and the matrix K is given by

$$K = \frac{1}{|\xi|^2 (P+Q)} \begin{pmatrix} \xi^* P + \xi Q & -\mu R^* e^{-i\delta} \\ \mu R e^{i\delta} & \xi P + \xi^* Q \end{pmatrix},$$

with $\xi = i\mu/2$, $N = \sqrt{k_s^2 - \mu^2}$, $P = h_{11} h_{11}^*$, $Q = h_{12} h_{12}^*$, $R = -ie^{-i\delta} h_{11}^* h_{12}$, $h_{11} = i(C_1 e^B - C_2 e^{-B}) e^{-i\delta/2}$, $h_{12} = (C_1 e^{-B} - C_2 e^B) e^{i\delta/2}$, $C_1 = \sqrt{i\,(A_s^2 k_s - N)/2}$, $C_2 = \sqrt{i\,(A_s^2 k_s + N)/2}$, and $B = -iN\,(x + i\mu t)/2$.

Akhmediev breathers. When $|\mu| < k_s$ we obtain the Akhmediev breathers in Eq. (2) with the parameters

$$P = k_s \cosh\theta - N \sinh\theta - \mu\cos\phi,\ Q = k_s \cosh\theta + N \sinh\theta - \mu\cos\phi,\ R = \mu\cosh\theta - k_s \cos\phi - iN \sin\phi, \tag{3}$$

where $\theta = \mu N t$ and $\phi = -Nx$. The above result reveals that the solutions in Eq. (2) and (3) is spatial periodic denoted by $2\pi/N$, and aperiodic in the temporal variable, as shown in Fig. 1 (a)-(c). This process can also be seen as the spatial manifestation of Fermi-Pasta-Ulam recurrence realized by the magnetization dynamics. The spatial periodic distribution of magnetization shows that the component m_3 has two peaks and two valley in each unit distribution. The magnetic Akhmediev breathers in Eqs. (2) and (3) in fact denotes the instability process of spin wave background. A periodic magnon exchange occurs between the magnetic soliton and the spin wave background. It should be noted that the magnetic soliton will lose this character on the ground state background. It is worth mentioning that the interaction between spin wave and magnetic soliton causes this very interesting phenomenon.

Kuznetsov-Ma soliton. Under the condition $|\mu| > A_s k_s$ we obtain the magnetic Kuznetsov-Ma soliton solution of Eq. (2) with the following parameters

$$P = \mu\cosh\theta - k_s\cos\phi - \zeta\sin\phi,\ Q = \mu\cosh\theta - k_s\cos\phi + \zeta\sin\phi,\ R = k_s\cosh\theta + i\zeta\sinh\theta - \mu\cos\phi, \tag{4}$$

where $\zeta = \sqrt{\mu^2 - k_s^2}$, $\theta = \zeta x$, and $\phi = \mu\zeta t$. The main characteristic properties of Kuznetsov-Ma soliton is spatially aperiodic and temporally periodic. Similar to the above discussion the component m_3 shows two peaks and two valley in each periodic distribution.

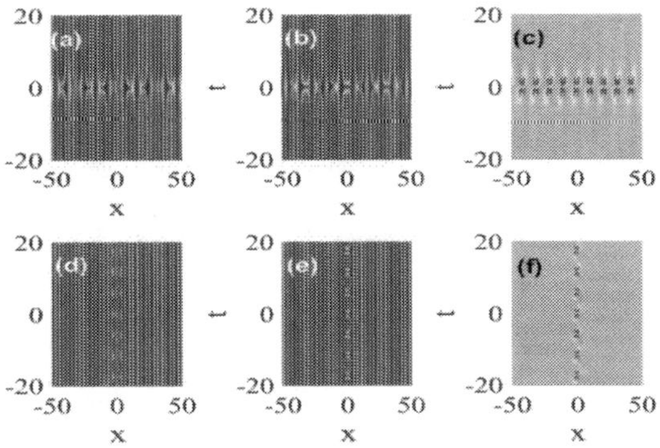

Figure 1. Evolution of Akhmediev breathers and Kuznetsov-Ma soliton for magnetization $\mathbf{m} = (m_1, m_2, m_3)$ in Eqs. (2), (3) and (4). Parameters are given as follows: $k_s = 1$, $\mu = 0.8$ for (a)-(c) and $\mu = 1.3$ for (d)-(f), respectively.

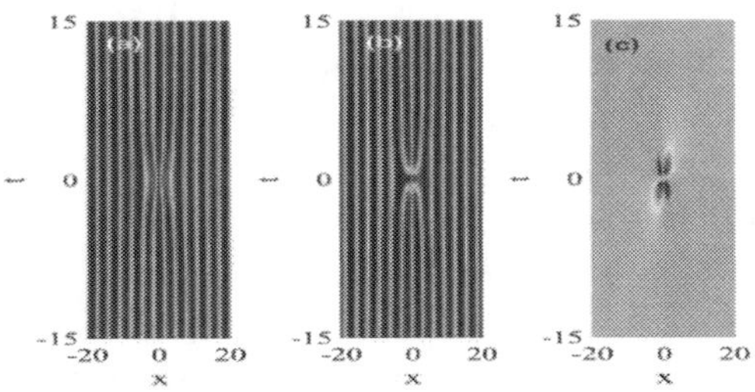

Figure 2. Evolution of rogue wave for magnetization $\mathbf{m} = (m_1, m_2, m_3)$ in the limit processes $\mu \to k_s$ with $k_s = 1$.

The illustration of magnetic Kuznetsov-Ma soliton is depicted in Fig. 1 (e)-(f), which shows that the soliton is trapped in space by spin wave background. The spin wave can affect the propagation velocity of magnetic soliton, which denotes the transfer of spin angular momentum from spin wave background to a dynamic soliton called magnonic spin-transfer torque [24]. The the magnon density distribution attains the maximum value 1 at $x \to \pm\infty$. Different from the Akhmediev breathers, the magnetic Kuznetsov-Ma soliton in Eqs. (2) and (4) expresses the localized periodic magnon exchange, which takes the temporal periodic evolution. Also, the high magnon density shows the temporal periodicity along the propagation direction of soliton.

Rogue waves. The above discussion shows that the condition $|\mu| = k_s$ forms a critical point which divides the modulation instability process ($|\mu| < k_s$) and the periodization process ($|\mu| > k_s$). It leads to the different physical behavior how the breather character depends strongly on the modulation parameter. There is two different asymptotic behavior in the limit processes $|\mu| \to (k_s)^-$ and $(k_s)^+$, respectively. The former case demonstrates a spatial periodic process of a magnetic soliton forming the petal with four pieces. The other case shows a localized process of the spin-wave background. In the limit case of $|\mu| \to k_s$, we get the novel magnetic rogue wave

$$m_+ = -e^{ik_s x}\left(1 - \left(8x^2k_s^2 - i4xk_s\left(F_1 - 2\right)\right)/F_1^2\right), m_3 = \pm 8txk_s^3/F_1^2, \tag{5}$$

where $F_1 = 1+t^2k_s^4+x^2k_s^2$. The component m_3 is characterized by the antisymmetric distribution of two peaks and two valleys, as shown in Fig. 2. The above results show that there exist

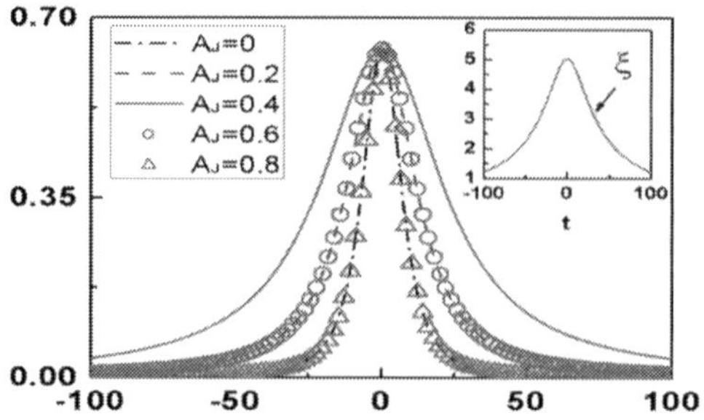

Figure 3. The nonuniform exchange of magnons between rogue wave and background for the different spin current. The inset figure denotes the maximal accumulation (or dissipation) process for the critical current value $A_J = 2k_c$. Other parameters are $A_c = 0.2$ and $k_c = 0.2$.

two processes of the formation of the magnetic rogue wave: one is the localized process of the spin wave background, and the other is the reduction process of the periodization of the magnetic bright soliton. The magnetic rogue wave is exhibited by the strong temporal and spatial localization of the magnon exchange and high magnon density.

3. Rogue waves with high magnon density distribution

For the perpendicular anisotropic ferromagnetic nanowire, it is reasonable to introduce q replacing the components of normalized magnetization [1], i.e., $q \equiv m_x + im_y$ and $m_z^2 = 1 - |q|^2$. Under the long-wavelength approximation [1], Eq. (1) becomes the integrable nonlinear Schrödinger equation

$$i\frac{\partial q}{\partial t} = \frac{\partial^2 q}{\partial x^2} + \frac{1}{2}q\left|q\right|^2 + iA_J\frac{\partial q}{\partial x} - \omega_0 q, \tag{6}$$

where $\omega_0 = 1 + h_{ext}/(h_k - 4\pi)$. In this case, the spin wave takes the form $q = A_c e^{-i(k_c x - \omega_c t)}$ with ω_c and k_c being the dimensionless frequency and wave number, respectively. By employing Darboux transformation [26] we get the novel magnetic rogue wave

$$q = A_c e^{i\varphi}\left[\frac{4\left(1 - itA_c^2\right)}{t^2 A_c^2 \eta + 2txA_c^2\kappa + \varepsilon} - 1\right], \tag{7}$$

where $\eta = A_J^2 + A_c^2 + 4k_c^2 - 4A_J k_c$, $\kappa = A_J - 2k_c$, $\varphi = k_c x - \omega_c t$, and $\varepsilon = 1 + x^2 A_c^2$. It is obvious that Eq. (7) shows the typical rogue wave feature that the magnons accumulated from spin wave background converge a single hump with the critical amplitude $A_Q = 3A_c$ with the high magnon density peak $|q|^2 = 9A_c^2$. It implies that the localization wave is captured completely at $x = 0$ and $t = 0$ by spin wave background.

In order to clarify the formation mechanism of rogue wave we consider the magnon density distribution $\rho \equiv \left|Q_1(x,t)\right|^2 - \left|Q_1(x = \pm\infty, t)\right|^2$ in a magnetic rogue wave, which takes the form

$$\rho = 8A_c^2\frac{\Gamma_1 - \Gamma_2}{\left(\Gamma_1 + \Gamma_2\right)^2}, \tag{8}$$

where $\Gamma_1 = 1 + t^2 A_c^4$ and $\Gamma_2 = A_c^2\left(x + t\left(A_J - 2k_c\right)\right)^2$. From Eq. (8) we find the integral $\int_{-\infty}^{+\infty}\rho(x,t)dx = 0$ and it shows that the loss of magnons in background completely transfer to hump. The generation of rogue wave is mainly arose from the gathering energy and magnons from the background toward to its central part, and the loss of magnons in background completely transfer to the hump part. On the other hand, it is interesting to show how rogue wave gather magnons and energy toward to its central part from the background. This can be explained by the quantity $\delta(x,t) \equiv \lim_{l_Q \to \pm\infty} |Q_1(x,t) - Q_1(x = l_Q, t)|^2$. With Eq. (7) we obtain

$$\delta(x,t) = 16A_c^2 \frac{\Gamma_1}{(\Gamma_1 + \Gamma_2)^2}, \tag{9}$$

which denotes the nonuniform exchange of magnons between rogue wave and background for the different spin current as shown in Fig. 3. From Eq. (9) we find the spin current can control the accumulation and dissipation rate of magnons, and there is a critical current condition, i.e., $A_{Jc} = 2k_c$. Below the critical current, the magnons exchange decreases with the increasing current term A_J. However, the magnons exchange is accelerated with the increasing current above the critical value. The roles of spin-transfer torque are completely opposite for the cases below and above the critical current which is shown in Fig. (3). When $A_J = 2k_c$, the time of magnons accumulation (or dissipation) attains its maximum. As shown the inset figure of Fig. (3), i.e., the integral $\xi(x,t) = \int_{-\infty}^{+\infty} \delta(x,t)\, dx = 8\pi A_c/(1 + t^2 A_c^4)^{1/2}$, the magnons in background accumulate to the central part when $t < 0$. It leads to the generation of a hump with two grooves on the background along the space direction and the critical peak of the hump can occur at $t = 0$. In contrast, when $t > 0$, the magnons in the hump start to dissipate into the background so that the hump gradually decay. The magnetic rogue wave disappears ultimately, and it verifies the rogue wave is only one oscillation in temporal localization and displays a unstable dynamic behavior.

4. Conclusions

In summary, we investigate the novel dynamics of magnetization in a ferromagnet excited by the All-Magnonic spin-transfer torque with the developed Darboux transformation. As an example, we obtain the exact expressions of Akhmediev breathers solution, Kuznetsov-Ma soliton and rogue waves. These results can be useful for the exploration of novel nonlinear excitation in Bosonic and fermionic ferromagnet.

5. Acknowledgments

This work was supported by the Key Project of Scientific and Technological Research in Hebei Province, China (Grant No. ZD2015133), the NKRDP under grants Nos. 2016YFA0301500, NSFC under grants Nos. 11434015, 61227902, 61378017, KZ201610005011, SKLQOQOD under grants No. KF201403, SPRPCAS under grants No. XDB01020300, XDB21030300.

6. References

[1] Kosevich A M, Ivanov B A, and Kovalev A S 1990 *Phys. Rep.* **194** 117
[2] Mikeska H J and Steiner M 1991 *Adv. Phys.* **40** 191
[3] He P B, Xie X C and Liu W M 2005 *Phys. Rev. B* **72,** 172411
[4] Tretiakov O A, Liu Y, and Abanov A 2012 *Phys. Rev. Lett.* **108** 247201
[5] Allwood D A, Xiong G, Faulkner C C, Atkinson D, Petit D, and Cowburn R P 2005 *Science* **309** 1688
[6] Parkin S S P, Hayashi M, and Thomas L 2008 *Science* **320** 190
[7] Hayashi M, Thomas L, Moriya R, Rettner C, and Parkin S S P 2008 *Science* **320** 209
[8] Jiang X, Thomas L, Moriya R, Hayashi M, Bergman B, Rettner C, and Parkin S S P 2010 *Nat. Commun.* **1** 25
[9] Hayashi M, Thomas L, Rettner C, Moriya R, and Parkin S S P 2007 *Nat. Phys.* **3** 21
[10] Thomas L, Moriya R, Rettner C, and Parkin S S P 2010 *Science* **330** 1810
[11] Schryer N L and Walker L R 1974 *J. Appl. Phys.* **45** 5406
[12] Gilbert T L 2004 *IEEE Trans. Magn.* **40** 3443
[13] Slonczewski M A, Magnetic Domain Walls in Bubble Materials (*Academic, New York,* 1979).
[14] Wang X R, Yan P, Lu J, and He C 2009 *Ann. Phys. (N. Y.)* **324** 1815
[15] Tatara G and Kohno H 2004 *Phys. Rev. Lett.* **92** 086601

IOP Conf. Series: Journal of Physics: Conf. Series **827** (2017) 012002 doi:10.1088/1742-6596/827/1/012002

[16] Li Z and Zhang S 2004 *Phys. Rev. Lett.* **92** 207203
[17] Saitoh E, Miyajima H, Yamaoka T, and Tatara G 2004 *Nature* (London) **432** 203
[18] Jiang X, Wang R, Shelby R M, Macfarlane R M, Bank S R, Harris J S, and Parkin S S P 2005 *Phys. Rev. Lett.* **94** 056601
[19] F. Freimuth, S. Blugel, and Y. Mokrousov, Phys. Rev. B **90,** 174423 (2014).
[20] Santos T S, Lee J S, Migdal P, Lekshmi I C, Satpati B, and Moodera J S 2007 *Phys. Rev. Lett.* **98** 016601
[21] Kasai S, Nakatani Y, Kobayashi K, Kohno H, and Ono T 2006 *Phys. Rev. Lett.* **97** 107204
[22] Sugimoto S, Fukuma Y, Kasai S, Kimura T, Barman A, and Otani Y 2011 *Phys. Rev. Lett.* **106** 197203
[23] Moriya R, Thomas L, Hayashi M, Bazaliy Y B, Rettner C, and Parkin S S P 2008 *Nat Phys* **4** 368
[24] Yan Y, Wang X S, Wang X R 2011 *Phys. Rev. Lett.* **107** 177207
[25] Li Z D, Li Q Y, Xu T F, and He P B 2016 *Phys. Rev. E* **94** 042202
[26] Zhao F, Li Z D, Li Q Y, Wen L, Fu G S, and Liu W M 2012 *Ann. Phys. (New York)* **327** 2085

Magnetic crossover realized by electrical methods

Hua Hao[1,2], **XiaoHong Zheng**[1,2], **Ting Jia**[1,2] **and Zhi Zeng**[1,2,3,*]

[1] Key Laboratory of Materials Physics, Institute of Solid State Physics, Chinese Academy of Sciences, Hefei 230031, China, [2] University of Science and Technology of China, Hefei, 230026, China, [3] Beijing Computational Science Research Center, Beijing 100084, China

E-mail: zzeng@theory.issp.ac.cn

Abstract. Electrical control of molecular-scale magnetisms is a greatly important topic in molecular spintronics. This enables the ultimate-limit manipulation of magnetisms, and has the potential to revolutionize computer technologies. Two mechanisms related to electrical control of molecular-scale magnetisms are concerned in this review. One is the magnetic crossover realized by the electrostatic Stark effect. The other is the magnetic crossover induced by the electron tunneling through the lowest unoccupied molecular orbital under bias voltages or one added electron under gate voltages.

1. Introduction

Over the past few decades, traditional transistors have shrunk dramatically and their dimensions are now less than 10 nm in research [1, 2]. Further miniaturization of transistors is however greatly challenging, due to technique limitations and lack of fundamental transport mechanisms [3]. On this issue, it is worth noting that organic molecules, with sizes of ca. 1 nm, can be synthesized in bulk while accomplishing a variety of electronic tasks, such as rectification, memory, and switching. These materials have the potential to replace some traditional solid-state device counterparts in the future, which gives birth to a new field called *molecular electronics* [4, 5, 6, 7]. The molecular electronics is conceptually fabricating electrical circuits using properties of individual organic molecules. Such circuits have three notable advantages: i) they can be manufactured with room-temperature low-cost chemical methods, instead of expensive high-temperature solid-state growth (e.g. molecular beam epitaxy) and pattering (lithography) techniques; ii) the extremely reduced size enables higher capacity and faster performance, and the limit of conventional silicon circuit integration can be surpassed; iii) the diversity of molecular structures probably brings some novel effects that are not accessible using traditional materials or approaches.

Among organic molecules, the ones with magnetic ions are exceptional objects for electronic devices, since the electrical resistance of devices with individual magnetic organic molecules can be modified by changing the magnetic texture. This has been effectively demonstrated by the giant magneto-resistance (GMR) effect in magnetic multilayers [8, 9]. The superiority of magnetic organic molecules over other organic molecules for electronic devices is that the energy scale relevant for manipulating the spin is orders of magnitude smaller than that involved in manipulating the electron charge in standard transistors. This can translate to molecular devices exhibiting ultra-low power consumption and high speed. Moreover, the powerless non-volatility becomes possible solely in magnetic materials with high Curie temperature. Therefore, an important branch is evoked and called as *Molecular Spintronics* [10, 11, 12], in which spin effects on electron transport properties are studied and molecular devices are proposed and designed using magnetic organic molecules.

As indicated above, in molecular spintronics the function of a device (e.g. rectification, memory, switching, etc.) is designed to be relied on magnetic textures of magnetic molecules. Thus it is most important to change the magnetic texture of a magnetic molecule in this branch. Moreover, for electronic applications it is desirable to accomplish changes of magnetic textures simply by applying an electric field. This would enable the ultimate-limit manipulation of magnetic textures and open a new pathway to spin-based quantum information processing [13], which have the potential to revolutionize computer technologies. Therefore, we will focus on two mechanisms related to the electrically-induced magnetic crossover in this review. One is based on the effect of electrostatic fields or the Stark effect. The other is closely associated with the electron tunneling through the lowest unoccupied molecular orbital (LUMO) under bias voltages or one added electron on magnetic molecules under gate voltages. Greater details are in the following.

2. Magnetic crossover achieved by electrostatic Stark effect

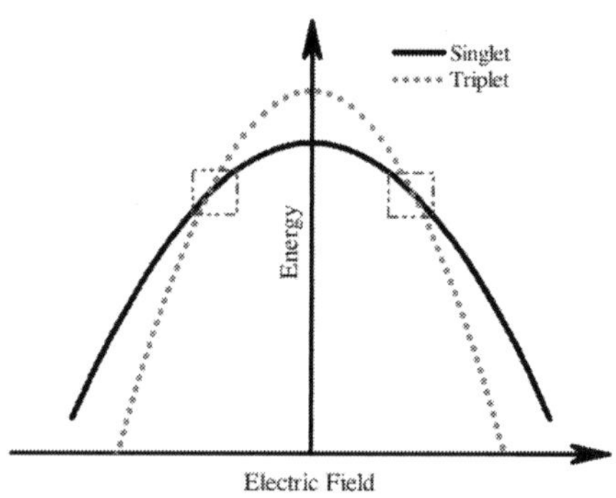

Figure 1. (color online). Schematic show for the electric-field induced spin crossover from singlet to triplet based on the second-order Stark effect. The squares mark the crossover points.

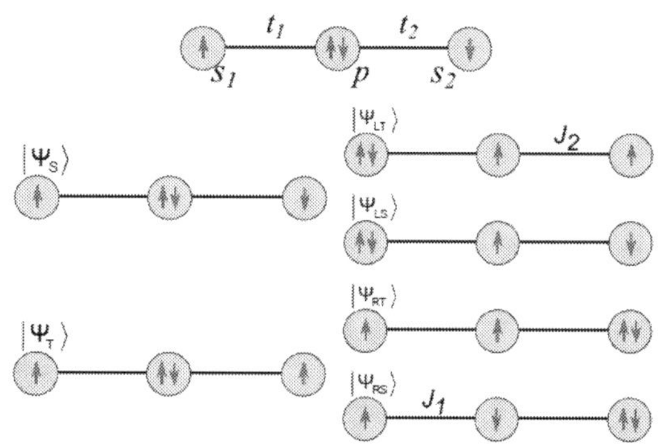

Figure 2. (color online). Schematic illustration of the three-center four-electron model for super-exchange. Reproduced with permission from Ref [15]. Copyright © 2009 Nature Publishing Group.

Diefenbach and Kim [14] firstly proposed that the Stark effect can be concerned to accomplish the spin-state crossover by an electric field. In their opinion, an external electrical field perturbs the electronic structure of a molecule. The effect of such a perturbation can be simply associated with the polarizability of the molecule and hence with the Stark effect. For a molecule with no permanent electric dipole, the linear (first-order) Stark effect vanishes, and the correction to the energy is given by the second-order Stark effect, which is proportional to the polarizability α of the molecule and to the square of the electric field $\mathbf{E}$ strength, i.e. $E_{el} = -\frac{1}{2}\alpha\mathbf{E}^2$. This relation applies to the ground spin states as well as to excited spin states, whereby each spin state responds differently to the perturbation and the spin-state crossover could be realized by an external electric field. As an example, their first-principles studies demonstrate that the ground state of the complex $Ba_2(C_6H_6)$ is changed from a nonmagnetic singlet to a magnetic triplet by an uniform external electric field, as shown by Fig. 1. This result lays the basis for the electrically-controlled manipulation of magnetism by means of the Stark effect. In addition, they suggest that a good candidate for a molecular magnetic switch must process the following conditions: i) the polarizability of the molecule must be high, and correspondingly, the gap between HOMO (highest occupied molecular orbital) and LUMO must be small, such that the electronic structure of the molecule is easily and distinctively perturbed by an external electric field; ii) the molecule must possess a nonmagnetic ground state (GS) with a magnetic excited state (ES) in close proximity (a very low GS-ES gap).

Similarly based on the Stark effect, Sanvito and co-workers [15] demonstrate that the sign of the exchange coupling between the two magnetic centers can be altered by an electric field. In their work, a simple model for super-exchange shown in Fig. 2 is used to illustrate the effect of an electric field on the sign of the exchange coupling between the two magnetic centers. The dependence of the energy gain on the electric field is accordingly depicted by

$$\Delta E_{\mathrm{GS}}^{\mathrm{S/T}} \sim \Delta E_{\mathrm{GS}}^{\mathrm{S/T}}(0) + \frac{1}{2a_0} p_{\mathrm{GS}}^{\mathrm{S/T}} V + \frac{1}{4a_0} \alpha_{\mathrm{GS}}^{\mathrm{S/T}} V^2, \tag{1}$$

and the electrical dipole moment and the polarizability are respectively defined as

$$p_{\mathrm{GS}}^{\mathrm{S/T}} = \frac{ea_0}{4} \left(\frac{t_1^2}{(\delta_{12}^{\pm})^2} + \frac{t_2^2}{(\delta_{21}^{\pm})^2} \right), \tag{2}$$

$$\alpha_{\mathrm{GS}}^{\mathrm{S/T}} = \frac{e^2 a_0}{4} \left(\frac{t_1^2}{(\delta_{12}^{\pm})^3} + \frac{t_2^2}{(\delta_{21}^{\pm})^3} \right). \tag{3}$$

Here, a_0 is the distance between the magnetic centers and the bridge, and the quantities $\delta_{ij}^{\pm} = E_i \pm J_j$ are the energies of the four excited states $|\Psi_{\alpha\beta}\rangle$ measured with respect to the energy of $|\Psi_S\rangle$ or $|\Psi_T\rangle$. E_i equals to $\Delta_{s_i} + U_{s_i} - U_p$, with U_{s_i} and U_p being the Coulomb charging energies of the s_i magnetic center and the bridge; Δ_{s_i} being the difference of their on-site energies. The four excited states $|\Psi_{\alpha\beta}\rangle$ depend on whether the hopping is to the left or to the right center (α=L, R), and whether the final molecular state is in an overall singlet or triplet state (β=S, T). The hopping integrals t, from the bridge to the magnetic center, differ for the two centers (i.e. t_1 and t_2), since this model is assumed to be generally asymmetric. As an example, their density-functional studies clearly show that a crossover of the magnetic coupling is accomplished by an uniform external electric field in the complex $MeO{-}CoCp_2{-}Ac{-}CoCp_2$ with a large permanent electric dipole. Moreover, they show that the critical field for the magnetic crossover can be engineered by using an appropriate choice of substituents to attach to the basic magnetic molecule. Such a strategy for the synthesis seems much simpler than the one needed for making the molecules with large electron density necessary for a crossover driven by the second-order Stark shift.

The above two works prove that electrical manipulations of magnetic textures can be realized by means of the Stark effect. However, such magnetic crossovers are just achieved in isolated molecules. In molecular devices or molecular junctions, the performance of the Stark effect is possibly different from that in isolated molecules, due to the presence of the metallic electrodes and the coupling of the molecule with the electrodes. Particularly, the condition for magnetic crossovers in molecular junctions may be greatly different from that in isolated molecules. To explicitly clarify the effect of the electrodes, we build a model junction [16], where a single magnetic molecule Fe_2(acpypentO)$(NCO)_3$ is sandwiched between two Au(100) nanoelectrodes [see Fig. 3 (a)]. Our investigations by density-functional theory plus the Keldysh nonequilibrium Green's function method demonstrate that the energy of this system is almost symmetrically raised as the bias voltage increases, which is displayed in Fig. 3 (b). This indicates a negative effective polarizability of the molecular junction, opposite to the positive polarizability of an isolated molecule. More importantly, there are some energy crosspoints between the ferromagnetic (FM) configuration and the antiferromagnetic (AFM) configuration around 0.4

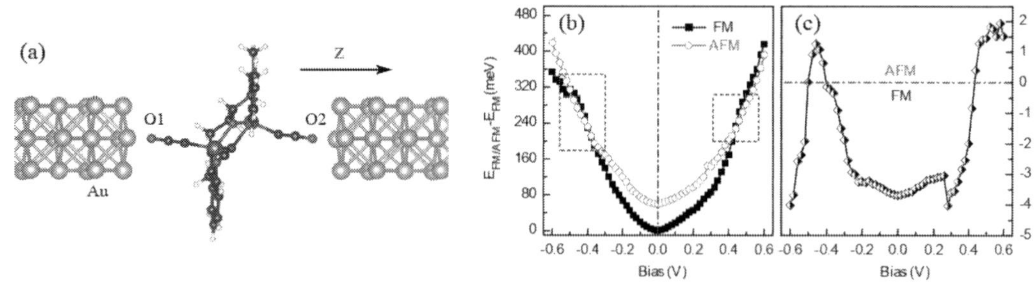

Figure 3. (color online). (a) the model molecular junction of Fe_2(acpypentO)$(NCO)_3$; (b) the relative energy $[E_{\mathrm{FM/AFM}} - E_{\mathrm{FM}}(V_B = 0)]$ as the function of the bias voltage; (c) the magnetic-coupling constant $[J_{\mathrm{FM/AFM}} = (E_{\mathrm{FM}} - E_{\mathrm{AFM}})/S^2_{\mathtt{max}}]$ as the function of the bias voltage. $S_{\mathtt{max}}$ is the maximum total spin of the molecule. Reproduced with permission from Ref [16]. Copyright © 2012 American Physical Society.

V [see Fig. 3 (c)] because of $\alpha_{\mathtt{FM}} < \alpha_{\mathtt{AF}} < 0$. In other words, the magnetic crossover from FM to AFM is realized by applying a certain bias voltage in our built molecular junction. On the contrary, in the isolated Fe_2(acpypentO)$(NCO)_3$ molecule no energy crosspoints are observed between the ground-state (GS) FM configuration and the excited-state (ES) AFM configuration due to $\alpha_{\mathtt{FM}} > \alpha_{\mathtt{AF}} > 0$. From these results, the electrodes and the molecule-electrode couplings indeed have great impact on the Stark effect in a molecular junction, and the sign of the effective polarizability in molecular junctions can be opposite to that in the isolated molecules. Moreover, the crossover from the GS magnetic configuration to the ES magnetic configuration is never observed when electric fields are applied to isolated molecules of $|\alpha_{\mathtt{GS}}| > |\alpha_{\mathtt{ES}}|$, but it can be realized if electric fields (or bias voltages) are applied to molecular junctions of these molecules sandwiched between two electrodes.

Finally, it is necessary to point out that the prediction of a magnetic crossover is just based on the relative energies of the two magnetic textures in the above description. In practice, the magnetic crossover also depends on the transition probabilities from one magnetic texture to another. An indication for this probability is related to the off-diagonal elements of the spin-orbital coupling matrix [14] or a non-spin-conversing interaction such as hyperfine interaction with nuclear spins [15].

3. Magnetic transition associated with one added electron on molecules

Recently, three well-established experiments report that the transition between low spin and high spin can be realized just by applying bias voltages to molecular junctions with an individual SCO molecule sandwiched between two metallic surfaces [17, 18, 19], which is a great advance in molecular spintronics. To explain these bias-induced spin transitions from low spin to high spin, several possible mechanisms have been suggested. For example, local resistive heating due to inelastic scattering of tunneling electrons or Joule effect may trigger this spin transition [17, 18]; the stability of the two spin states may be influenced by Stark effect [14, 15, 16] or other electrostatic field effects; electron-induced excited spin state trapping may play a key role in this spin transition, analogous to light-induced excited spin state trapping [19, 20]. As a matter of fact, the spin transition from low spin to high spin is only observed by applying a certain negative bias voltage in the two experiments [17, 19], but not by any positive bias voltage. Due to this feature, the Joule-effect mechanism could not be the underlying cause of these two bias-induced spin transitions, because inelastic scattering of tunneling electrons is present at both bias polarities. On the other hand, electrostatic field effects could also be hardly responsible for these bias-induced spin transitions, which has been directly confirmed by

Gopakumar *et al* [19]. For the last possibility, its details have been far from clear, such as the information of excited states, its definite effect on the spin transition, and the explanation for the dependence of spin transition on the bias polarity, etc.

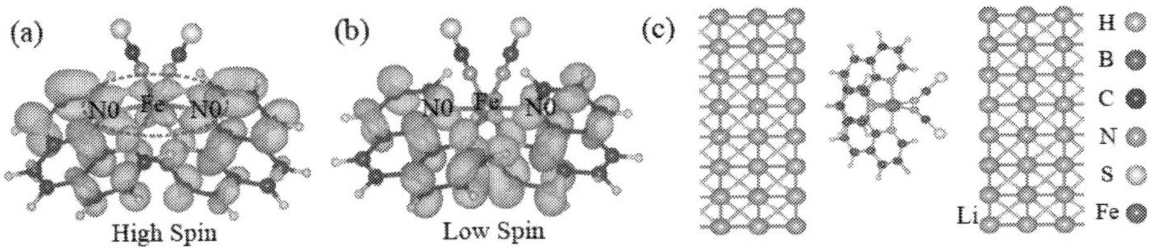

Figure 4. (color online). (a) and (b) the real-space distributions of the added electron in the high-spin and low-spin phases of $Fe(phen)_2(NCS)_2$. The circle marks the *p*-*d* hybridization between N0 and Fe. (c) the molecular junction of $Fe(phen)_2(NCS)_2$ sandwiched between two metallic leads. The 'phen' and 'NCS' groups are respectively connected to the left and right electrodes. Reproduced with permission from Ref [21].

Our latest work has explicitly revealed the mechanism for the bias-induced spin transition from low spin to high spin [21]. In this work, we propose that the bias-induced spin transition from low spin to high spin is associated with the electron tunneling through the LUMO of SCO molecules or one added electron on these molecules. To prove such a proposal, two steps are required: i) one added electron on the SCO molecule can accomplish the spin transition from low spin to high spin; ii) one electron can be added to the SCO molecule when a certain bias voltage is applied to a molecular device. Firstly, our density-functional studies show that the spin transition from low spin to high spin is indeed accomplished by adding one electron to the SCO molecule (e.g. $Fe(phen)_2(NCS)_2$). Moreover, by analyzing the related data, this one electron induced spin transition is ascribed to the respective occupations of a *p*-*d* bonding and a *p*-*d* non-bonding states in the high-spin and low-spin phases, as displayed in Fig. 4 (a) and (b). Secondly, our electron transport studies prove that one electron can be added to the central SCO3 molecule by applying negative bias voltages to our built molecular junction of an individual $Fe(phen)_2(NCS)_2$ [see Fig. 4(c)], but one added electron is hardly realized under positive bias voltages. That is to say, one added electron on the central SCO3 molecule is determined by the bias polarity in this junction. This fact is directly related to the inherently unequal couplings of the molecule with the two electrodes, indicated by that different groups of the molecule are contacted with the left and right electrodes. Here, it is pointed out that our built junction is similar to the experimental junction [17], and the left (right) electrode can be equivalent to the STM tip (substrate). Negative (positive) bias voltages are defined to raise the electrochemical potential of the left (right) electrode with respect to the right (left) electrode, which is equivalently the same with the definition of the bias polarity in this experiment as well. Based on such two-step studies, one can easily appreciate the spin transition from low spin to high spin induced by bias voltages, especially the dependence of spin transition on bias polarity in the experiment.

4. Conclusions

In this review for electrical control of molecular-scale magnetisms, two mechanisms are illustrated: one is based on the electrostatic Stark effect, the other is related to the electron tunneling through the LUMO under bias voltages or one added electron under gate voltages. For the first one, it is emphasized that the electrodes have great impact on the Stark effect in a molecular junction. As a consequence, the magnetic crossover of a molecule, absent (present) in an isolated situation, probably succeeds (fails) in a molecular device. As for the second one, the fundamental cause for the magnetic crossover is that different *p-d* states are occupied at different magnetic phases under bias or gate voltages. Based on this mechanism, one can well appreciate the spin transition dependent on the bias polarity. Moreover, it will also facilitates rational and efficient design of molecular devices (e. g. transistors, memories) using SCO molecules. Finally, although some advances have been made in exploring mechanisms related to electrically-induced magnetic transitions, it is still necessary to uncover more other related mechanisms, and particularly, it is more important to design molecular spin devices by taking full advantage of these mechanisms in experiments.

Acknowledgments
This work was supported by the National Science Foundation of China under Grant Nos. 11374301, 11574318, 11534012, and U1230202 (NSAF), the special Funds for Major State Basic Research Project of China (973) under Grant No. 2012CB933702 and Director Grants of CASHIPS. Our calculations were partly performed in Center for Computational Science of CASHIPS, the ScGrid of Supercomputing Center and Computer Net work Information Center of Chinese Academy of Science, and partly using Tianhe-2JK computing time award at the Beijing Computational Science Research Center (CSRC).

References

[1] Lortscher E 2013 *Nature Nanotech.* **8** 381
[2] Wilson L 2013 *International Technology Roadmap for Semiconductors (ITRS)* Semiconductor Industry Association
[3] Haedler A T, Kreger K, Issac A, Wittmann B, Kivala M, Hammer N, Kohler J, Schmidt H W and Hildner R 2015 *Nature* **523** 196
[4] Ratner M 2013 *Nature Nanotech.* **8** 378
[5] Bryce M R and Petty M C 1995 *Nature* **374** 771
[6] Tour J M, Reinerth W A, Jones L, Burgin T P, Zhou W C, Muller C J, Deshpande M R and Reed M A 1998 *Ann. N. Y. Acad. Sci.* **852** 197
[7] Petty M C, Bryce M R and Bloor D 1995 *An introduction to Molecular Electronics* (New York: Oxford Univesity Press)
[8] Baibich M N, Broto J M, Fert A, Nguyen Van Dau F, Petroff F, Etienne P, Creuzet G, Friederich A and Chazelas J 1988 *Phys. Rev. Lett.* **61** 2472
[9] Binasch G, Grünberg P, Saurenbach F and Zinn W 1989 *Phys. Rev.* B **39** 4828.
[10] Sanvito S and Rocha A R 2006 *J. Comput. Theor. Nanosci.* **3** 624
[11] Bogani L and Wernsdorfer W 2008 *Nature Mater.* **7** 179
[12] Kim W Y, Choi Y C, Min S K, Cho Y and Kim K S 2009 *Chem. Soc. Rev.* **38** 2319
[13] Loss D and DiVincenzo D P 1998 *Phys. Rev.* A **57** 120
[14] Diefenbach M and Kim K S 2007 *Angew Chem. Int. Ed.* **46** 7640
[15] Baadji N, Piacenza M, Tugsuz T, Sala F D, Maruccio G and Sanvito S 2009 *Nature Mater.* **8** 813
[16] Hao H, Zheng X H, Song L L, Wang R N and Zeng Z 2012 *Phys. Rev. Lett.* **108** 017202
[17] Miyamachi T, Gruber M, Davesne V, Bowen M, Boukari S, Joly L, Scheurer F, Rogez G, Yamada T K, Ohresser P, Beaurepaire E and Wulfhekel W 2012 *Nature Commun.* **3** 938
[18] Prins F, Monrabal-Capilla M, Osorio E A, Coronado E and van der Zant H S J 2011 *Adv. Mater.* **23** 1545
[19] Gopakumar T G, Matino F, Naggert H, Bannwarth A, Tuczek F and Berndt R 2012 *Angew. Chem. Int. Ed.* **51** 6262
[20] Hauser A 2004 *Top. Curr. Chem.* **234** 155
[21] Hao H, Jia T, Zheng X H and Zeng Z 2016 *Spin transitions induced by electron tunneling through lowest unoccupied orbital of spin crossover molecules* unpublished

IOP Conf. Series: Journal of Physics: Conf. Series **827** (2017) 012004 doi:10.1088/1742-6596/827/1/012004

Micromagnetic simulation of two-body magnetic nanoparticles

Fei Li[1], Jincheng Lu[1], Yu Yang[1], Xiaofeng Lu[1], Rujun Tang[1, 2] and Z. Z. Sun[1, 2, *]

[1]College of Physics, Optoelectronics and Energy, Soochow University, Suzhou, Jiangsu 215006, China

[2]Jiangsu Key Laboratory of Thin Films, Soochow University, Suzhou, Jiangsu 215006, China

[*]phzzsun@suda.edu.cn

Abstract. Field-induced magnetization dynamics was investigated in a system of two magnetic nanoparticles with uniaxial anisotropies and magnetostatic interaction. By using the micromagnetic simulation, ultralow switching field strength was found when the separation distance between the two particles reaches a critical small value on nanometer scale in the perpendicular configuration where the anisotropic axes of the two particles are perpendicular to the separation line. The switching field increases sharply when the separation is away from the critical distance. The same results were observed when varying the radius of particles. The micromagnetic results are consistent with the previous theoretical prediction where dipolar interaction between two single-domain magnetic particles was considered. Our present simulations offered further proofs and possibilities for the low-power applications of information storage as the two-body magnetic nanoparticles could be implemented as a composite information bit.

1. Introduction

Nanomagnetism has attracted wide attention for potential applications in information storage industry, such as hard disk technology, magnetic memory, and logic devices [1, 2]. The magnetization switching of Stoner particles which are believed to be advantageous for high-density information storage has been numerously studied and made significant progress in recent years, both theoretically and experimentally [3-14]. For example, an energy surface model was introduced into the micromagnetic simulation for a single-domain particle and a reversal field below the famous Stoner-Wohlforth limit was observed [15]. When the nanoparticles were fabricated in array patterns [6, 16-19] the magnetostatic interaction among particles plays an important role and will significantly affect the magnetization switching behaviors. Two-body Stoner particles system could be the simplest one to investigate the magnetostatic interaction on the magnetic switching [20-26]. The dipolar interaction between two particles can assist the switching and influence their hysteresis loops in the two-body system, which was numerically explored in Ref. [20, 23]. In the previous work [27], we have predicted that a critical switching field appears to be dramatically low, almost to the zero-field regime by omitting thermal fluctuation, with simply considering the magnetic dipolar interaction between the two particles. Our present work is desire to verify it with the more realistic micromagnetic simulation.

In this work, we have studied the magnetization switching of the two-body system and verified our previous results [27] by using the micromagnetic simulation OOMMF package [28]. An ultralow switching field strength (about 20mT) was found when the separation between the two Cobalt particles of 10nm-radius reaches a critical small value (about 23nm). Moreover, varying the radius of nanoparticles the same results were observed. The micromagnetic results verified our theoretical predictions [27], which shall be believed to offer further possibilities for the low-power applications of information storage.

2. Model

The dynamics of magnetization of each particle in a system of two magnetic nanoparticles can be described by the Landau-Lifshitz-Gilbert (LLG) equation [7, 27],

$$\frac{\partial \vec{M}_i}{\partial t} = -|\gamma| \vec{M}_i \times \vec{H}_{i,eff} + \frac{\alpha}{M_s}\left(\vec{M}_i \times \frac{\partial \vec{M}_i}{\partial t}\right). \quad (1)$$

Here $\vec{M}_i$ is the magnetization vector of the ith particle ($i = 1,2$). $M_s = |\vec{M}_i|$, γ and α are the saturation magnetization of each particle, the gyromagnetic ratio and the Gilbert damping coefficient, respectively. $\vec{H}_{i,eff}$ is the effective magnetic field acting on the ith particle. The material parameters are for Cobalt ($M_s = 1.4 \times 10^6\, A/m, A = 1.3 \times 10^{-11}\, J/m, K_i = 10^5\, J/m^3$). Damping coefficient $\alpha = 0.5$ is chosen bit larger for accelerating the simulation speed in our static simulation without loss of accuracy. Both particles are in spherical shapes with the same radius, and the mesh cell size is $1\text{nm} \times 1\text{nm} \times 1\text{nm}$. The two-body system consists of two identical magnetic nanoparticles with uniaxial anisotropies (see inset in Figure 1). According to our previous results [27] perpendicular configuration is a more promising candidate of a composite information bit, in which the connecting unit vector $\vec{n}$ between the centers of the two particles is perpendicular to the anisotropy easy-axes (EA) of each particle, referred to as PERP configuration.

3. Results and discussion

At the first step we considered the PERP configuration with particles of 10nm-radius and the varying separation d. The initial magnetization directions of the system were set to along the easy axis (x-direction) and the external fields were applied antiparallel to the initial magnetization directions [see inset in Figure 1(b)]. Figure 1(a) shows the magnetic hysteresis loops of magnetization for the varying separation distances ranging from 22nm to 120nm. The curves clearly show that the coherent rotation is satisfied with a large range of separations. Figure 1(b) shows the dependence of the switching fields on the varying separations. An ultralow switching field strength of about 17mT was observed for the separation chosen as $d = 23$nm. The corresponding magnetic vector distribution snapshots within particles are shown in Figure 2(b) which corresponds to the two locations D (no magnetization reversal), E (both particles reversed) in Figure 1(a) (red color). The behaviors for other larger separations are similar. For $d < 23nm$, the switching fields abruptly become large where two metastable states appear as shown in the hysteresis curve of $d = 22$nm in Figure 1(a) (black color). Two midst plateaus of zero total magnetization of the two particles can be clearly observed in the figure. The switching field for totally reversing the two particles is found at about 186mT. The corresponding magnetic vector distribution snapshots within particles for the metastable state are shown in Figure 2(a) which corresponds to the three locations A (no magnetization reversal), B (one particle reversed), C (both particles reversed) in Figure 1(a) (black color), respectively. It implies that one of two particles reverses its magnetization first, and then the other particle reverses. This phenomenon could be interpreted as the enhancement of the magnetostatic interaction with the very close distance between the two particles.

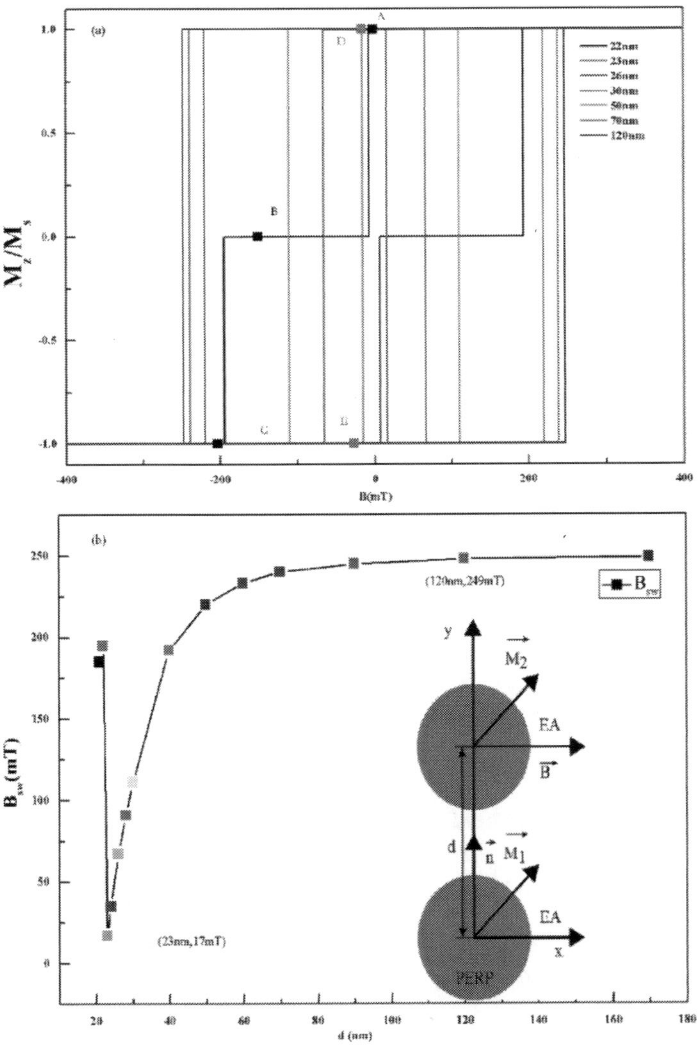

Figure 1. (Color online) (a) The hysteresis loops of magnetization for different separations d in the PERP configuration. (b) The critical switching field versus the separations d ranging from 21nm to 170nm when magnetic fields are applied along the x-direction shown in the lower right inset. The system parameters are $K_i = 10^5 J/m^3$, and $\alpha = 0.5$.

On the other hand, the switching field increases in saturation with the rising separations ranging from 23nm to 120nm. It approaches to a constant value of 250mT for larger separations. The value is closely equal to the switching field required for a single nanoparticle under the same parameters observed in our simulation. The increasing trend of the switching field is due to the competition between the magnetostatic interaction and the exchange interaction which are significant when the separations between the two particles are much closer. Nevertheless, both of them can be negligible when two bodies are separated far away. Thus, the far-away behavior of the two-body system will behave as a single particle.

Figure 3 reveals that the critical switching fields versus the separations for different radius of particles. As shown in the figure, the minimum switching fields can be found in certain separations that is 18nm, 23nm and 28nm corresponding to the radius of 8nm, 10nm and 12nm, respectively. These results are consistent with the previous theoretical prediction with the critical seperation distance $d_c = \left(\frac{\pi\mu_0 M_S^2}{2\pi K}\right)^{1/3} r$ [27]. Consequently, a ultralow magnetic field is sufficient to switch the magnetization in a system of two-body particles from initial state to the antiparallel state. More detailed physical significance about the above critical condition has been illustrated in Ref. [27].

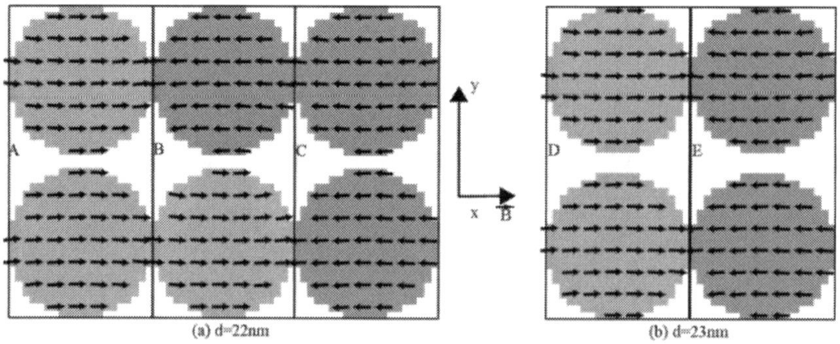

Figure 2. (Color online) The magnetic vector distribution snapshots inside the particles for (a) $d = 22\text{nm}$ and (b) $d = 23\text{nm}$. Each single snapshot from A to E corresponds to the marked locations in Figure 1(a). Orange and green color depict with magnetization reversal and without reversal, respectively.

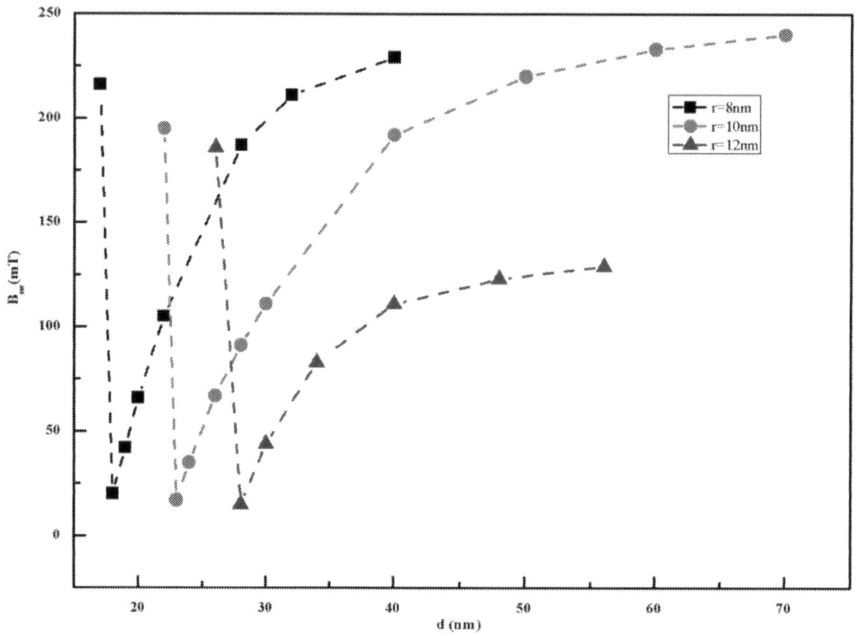

Figure 3. (Color online) The critical switching field versus the separations d for radius of 8nm, 10nm and 12nm of particles in PERP configuration

The present work has the advantages over the previous studies with taking into account the more realistic micromagnetic simulation case. In the simulation, the magnetostatic interaction is adopted in place of the simplified dipolar interaction in the previous analytical calculations. Thus the ultralow field switching behavior could be more realistic and be tested in real experiments. Additionally, we have also simulated the magnetization switching of the system with the initial magnetic states of two particles antiparallel to each other along the easy-axes. The results were similar which implies that the choices of the initial states will not affect much on the ultralow magnetization switching behaviors.

4. Summary

In conclusion, we have investigated magnetization switching of two-body magnetic nanoparticles by using micromagnetic simulation. Owing to the competition between magnetostatic interaction and magnetic exchange interaction, an ultralow magnetization switching field was found in the perpendicular configuration for a critical separation distance between two particles. A metastable state of the two particles was also observed below the critical distance. The switching fields increase with the increasing separations beyond the critical distance and finally reach a saturated value because the contribution of magnetostatic interaction and exchange interaction is getting weaker as the two particles separate far away. Moreover, varying the radius of particles shows the same behaviors. It should be pointed out that our simulation neglects the thermal fluctuations. The micromagnetic simulation results have verified our previous theoretical predictions and could offer more possibilities to new magnetic information storage applications.

References

[1] Hillebrands B and Ounadjela K 2002 and 2003 *Spin dynamics in confined magnetic structures I&II* (Berlin: Springer-Verlag)
[2] Hillebrands B and Thiaville A 2006 *Spin dynamics in confined magnetic structures III* (Berlin: Springer-Verlag)
[3] Stoner E C and Wohlfarth E P 1948 *Philos. Trans. R. Soc. London Ser.* **A240** 599
[4] Hiebert W K, Stankiewicz A and Freeman M R 1997 *Phys. Rev. Lett.* **79** 1134
[5] Stamps R L and Hillebrands B 1999 *Appl. Phys. Lett.* **75** 1143
Bauer M, Fassbender J, Hillebrands B and Stamps R L 2000 *Phys. Rev. B* **61** 3410
[6] Schumacher H W, Chappert C, Sousa R C, Freitas P P and Miltat J 2003 *Phys. Rev. Lett.* **90** 017204
[7] Sun Z Z and Wang X R 2005 *Phys. Rev. B* **71** 174430
Sun Z Z and Wang X R 2006 *Phys. Rev. B* **73** 092416
Sun Z Z and Wang X R 2006 *Phys. Rev. B* **74** 132401
[8] Sun Z Z and Wang X R 2006 *Phys. Rev. Lett.* **97** 077205
Wang X R, Yan P, Lu J and He C 2008 *Europhys. Lett.* **84** 27008
[9] Acremann Y, Back C H, Buess M, Portmann O, Vaterlaus A, Pescia D and Melchior H 2000 *Science* **290** 492
Acremann Y, Back C H, Buess M, Pescia D and Pokrovsky V 2001 *Appl. Phys. Lett.* **79** 2228
[10] Tsoi M, Jansen A G M, Bass J, Chiang W C, Seck M, Tsoi V and Wyder P1998 *Phys. Rev. Lett.* **80** 4281
Kiselev S I, Sankey J C, Krivorotov I N, Emley N C, Schoelkopf R J, Buhrman R A and Ralph D C 2003 *Nature* **425** 380
[11] Slonczewski J 1996 *J. Magn. Magn. Mater.* **159** L1
Berger L 1996 *Phys.Rev. B* **54** 9353
Bazaliy Y B, Jones B A and Zhang S C 1998 *Phys. Rev. B* **57** R3212
[12] Stiles M D and Zangwill A 2002 *Phys. Rev. B* **66** 014407
[13] Wang X R and Sun Z Z 2007 *Phys. Rev. Lett.* **98** 077201
[14] Sukhov A and Berakdar J 2009 *Phys. Rev. Lett.* **102** 057024
[15] Gao K Z, Boerner E D and Bertram H N 2003 *J. Appl. Phys.* **93** 6549
[16] Ntallis N and Efthiniadis K G 2015 *Computational Materials Science* **97** 42-47
[17] Hehn M, Ounadjela K, Bucher J P, Rousseaux F, Decanini D, Bartenlian B and Chappert 1996 *Science* **272** 1782
[18] Stamm C, Marty F, Vaterlaus A, Weich V, Egger S, Maier U, Ramsperger U, Fuhemann H and Pescia D 1998 *Science* **282** 449
[19] Sun S, Murray C B, Weller D, Folks L and Moser A 2000 *Science* **287** 1989
[20] Bertram H N and Mallinson 1969 *J. Appl. Phys.* **40** 1301
Bertram H N and Mallinson J C 1970 *J. Appl. Phys.* **41** 1102
[21] Chen W, Zhang S and Bertram H N 1992 *J. Appl. Phys.* **71** 5579

Lu J J, Liu M T, Kuo C C and Huang H L 1999 *J. Appl. Phys.* **85** 5558
Huang H L and Lu J J 1999 *Appl. Phys. Lett.* **75** 710
[22] Lyberatos A and Chantrell R W *J. Appl. Phys.* **73** 6501
[23] Xu C, Hui P M, Zhou J H and Li Z Y 2002 *J. Appl. Phys.* **91** 5957
Zhang L F, Xu C, Hui P M and Ma Y Q 2005 *J. Appl. Phys.* **97** 103912
[24] Pham H N, Dumitru I, Stancu A and Spinu L 2005 *J. Appl. Phys.* **97** 10P106
[25] Plamada A V, Cimposu D and Stancu A 2010 *Appl. Phys. Lett.* **96** 122505
[26] Sun Z Z, Lopez A and Schliemann 2011 *J. Appl. Phys.* **109** 102303
[27] The OOMMF code is available at http://math.nist.gov/oommf

Acknowledgments
We acknowledge the support for this work by the National Natural Science Foundation of China (NSFC Grant No. 11274236, No. 51202153), Research Fund for the Doctoral Program of Higher Education of China (RFDP Grant No. 20123201110003, No.20123201120028).

IOP Conf. Series: Journal of Physics: Conf. Series **827** (2017) 012005 doi:10.1088/1742-6596/827/1/012005

The collinear ↑↑↓↓ magnetism driven ferroelectricity in double-perovskite multiferroics

Ting Jia[1], **Zhi Zeng**[12], **and H. Q. Lin**[3]

[1]Key Laboratory of Materials Physics, Institute of Solid State Physics, Chinese Academy of Sciences, Hefei 230031, People's Republic of China

[2]*Department of Physics, University of Science and Technology of China, Hefei, 230026, China*

[3]*Beijing Computational Science Research Center, Beijing 100084, China*

E-mail: `Corresponding author: zzeng@theory.issp.ac.cn`

Abstract. The multiferroics of collinear ↑↑↓↓ magnetism driven ferroelectricity is one type of the magnetically driven ferroelectrics, which attracts much attentions due to their strong magnetoelectric coupling. Here, we summarize the recent progress in this multiferroics with double-perovskite crystal structure, besides Y_2CoMnO_6, Lu_2CoMnO_6, Y_2NiMnO_6 and In_2NiMnO_6 etc.. It is revealed that there are also many uncertainties in present research, making this field fulling of challenges and opportunities.

1. Introduction

Multiferroics, in which ferromagnetic and ferroelectric orders coexist and couple with each other, have attracted much attention due to their interacting fundamental physics and potential applications to memory and magnetic storage devices[1, 2]. Multiferroic materials are divided into two types by Khomskii in 2009[3]. Type-I multiferroics exhibit good ferromagnetism and ferroelectricity but weak magnetoelectric (ME) coupling, because the ferromagnetism and ferroelectricity have different sources and appear largely independently of one another. For the Type-II multiferroics, the polarization is usually much smaller and the transition temperature is relatively lower than type-I, but it exhibits strong ME coupling since the ferroelectricity is induced by magnetic order, which is important for the electric control of magnetism or vice versa[4]. There are two magnetic configurations as the source of ferroelectricity. One is an noncollinear or a spiral magnetic order, which is usually found in frustrated systems. The other is a collinear magnetic structure, typically an up-up-down-down (↑↑↓↓) magnetic structure, which causes an exchange striction and thus breaks inversion symmetry to get a polarization.

The ↑↑↓↓ magnetic structure can be divided into two classes according to the orientation of ↑↑↓↓ spin order. One is called E-type antiferromagnetic (AFM) state that the ↑↑↓↓ spin alignment along the in-plane [110] direction, as shown in Fig.1(a). This E-type AFM state has taken place in the orthorhombic $RMnO_3$ perovskites where R is a small rare earth, such as $HoMnO_3$ and $YMnO_3$[5]. The exchange striction effect here leads to the alternation of the long and short Mn-Mn bonds in the *ab* plane and meanwhile causes the oxygen ions to shift perpendicular to the Mn-Mn bonds, thus producing a polarization along the shift direction[6]. The other ↑↑↓↓ magnetic structure has been firstly found in Ca_3CoMnO_6[7, 8], where the ↑↑↓↓ spin order is along the Co-Mn chains (*c*-axis), calling as E*-type AFM

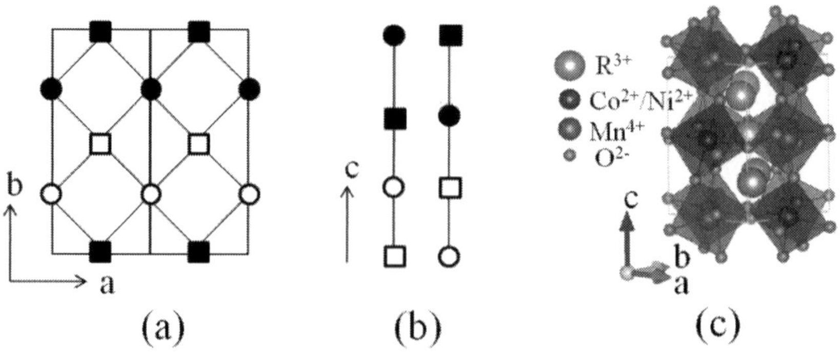

Figure 1. Schematic representation of spin configuration (a)E-type antiferromagnetic state of R_2NiMnO_6, squares and circles denote the Ni^{2+} and Mn^{4+} ions, respectively; (b)E*-type antiferromagnetic state of R_2CoMnO_6, squares and circles denote the Co^{2+} and Mn^{4+} ions, respectively. The up and down spins are represented by the white and black filling. (c)The monoclinic structure of R_2CoMn_6 or R_2NiMnO_6.

state(Fig.1(b)). The exchange striction in Ca_3CoMnO_6 arises from the stronger exchange interaction between adjacent metal ions with identical spin (next-nearest-neighbor (NNN) Co-Co and Mn-Mn) than that between adjacent metal ions with opposite spins (nearest-neighbor (NN) Co-Mn)[9]. Recently, this ↑↑↓↓-type multiferroics have been extended to double perovskites R_2CoMnO_6[10, 11, 12, 13, 14] and R_2NiMnO_6[15, 28], where the R_2CoMnO_6 belongs to E*-type (↑↑↓↓ along Co-Mn chains) while R_2NiMnO_6 belongs to E-type (↑↑↓↓ in Ni-Mn plans).

For these double perovskite multiferroics, the ferroelectricity is also mainly from the magnetically induced exchange striction effect. However, the origin of magnetic order and the polarize direction are still under debate. Then, we will summarize recent progress in this multiferroics.

2. R_2CoMnO_6

The structure of R_2CoMnO_6 is monoclinic lattice ($P2_1/n$). As shown in Fig. 1(c), the MnO_6 and CoO_6 octahedral are corner-sharing and alternate along all *a*, *b*, and *c* directions in double-perovskite. This structural characteristic is obviously different from that of Ca_3CoMnO_6, in which the MnO_6 and CoO_6 octahedral are face-sharing along *c* direction and $CoMnO_6$ chains are isolated by Ca ions. Thus, the NNN Co-Co or Mn-Mn distances in R_2CoMnO_6 are much larger (e.g. 7.41Å in Lu_2CoMnO_6 and 7.46Å in Y_2CoMnO_6) than those of Ca_3CoMnO_6 (5.29 Å). Hence, the exchange interactions of NNN Co-Co/Mn-Mn are greatly weakened and not comparable to those of the NN Co-Mn. Although the R_2CoMnO_6 belongs to E*-type multiferroics, the origin of ME properties should be different from that of Ca_3CoMnO_6.

Not mentioned the ME origin, the existence of E*-type AFM order and ferroelectricity in reported R_2CoMnO_6 is still controversial. As we know, R is a small rare earth in $RMnO_3$ with ↑↑↓↓-type magnetic structure. The reported ↑↑↓↓-type multiferroics of R_2CoMnO_6 also have a small rare earth R, such as Lu_2CoMnO_6[10, 11], Yb_2CoMnO_6[12], Y_2CoMnO_6[13] and Sm_2CoMnO_6[14]. The E*-type magnetic order and multiferroic behavior of Y_2CoMnO_6 and Sm_2CoMnO_6 are still controversial. Only the Lu_2CoMnO_6 and Yb_2CoMnO_6 are clarified to be a ↑↑↓↓ magnetically induced ferroelectricity by neutron diffraction[11, 12]. We will take the most reported Y_2CoMnO_6 and Lu_2CoMnO_6 as examples to demonstrate the progress in this series.

For Y_2CoMnO_6, the E*-type magnetism driven ferroelectricity has been firstly reported by

Sharma et al.[13]. And they claim that Y_2CoMnO_6 undergoes the ME transition at relatively higher temperature 80 K. However, such ground state contradicts the reports before, in which the magnetic ground state of Y_2CoMnO_6 is ferromagnetic (FM) ordering[17, 18, 19]. This controversy attracts much attention from theory and experiment. The recent experimental study of Y_2CoMnO_6 nearly confirms the FM ground state[20, 21], although E*-type magnetism is obtained from the first-principles calculations[22, 23]. The results from the first-principles calculations observe that besides the E*-type AFM ground state, the metastable FM state is only a little higher in energy. Moreover, the ground state depends on the Hubbard U values, for which the FM state could be the ground state at $U > 1.8$ eV while E*-type magnetism is the ground state at $U < 1.8$ eV[22]. In addition to the uncertainty on the ground state, the direction of electric polarization based on the presumed ME state is also controversial. Some suggest that the polarization is along c-axis (Fig. 2(a)), where the ferroelectricity is from the exchange-striction induced non-centrosymmetric structural distortions of inequivalent Co-Mn bonds along c-axis[13, 23]. Others think that the electric polarization aligns along the b-axis (Fig. 2(b)), where the ferroelectricity is generated by the net displacements of O ions along the b-axis[22]. Consequently, there are many uncertainties in the research of Y_2CoMnO_6, especially the existence of ME preterites should be clarified first.

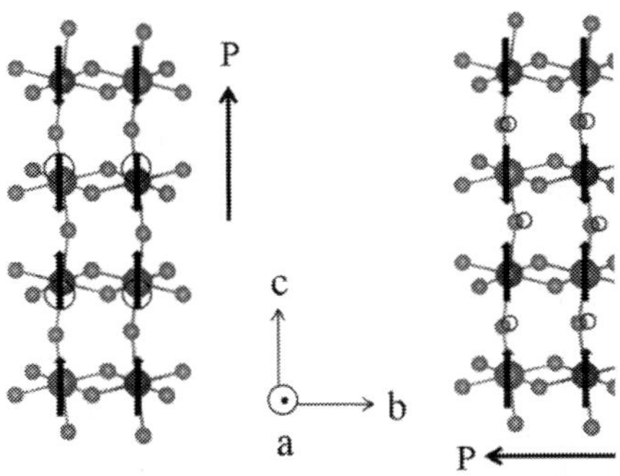

Figure 2. [22]The exchange striction induced displacements of (a) Co and Mn ions along c-axis and (b) O ions along b-axis, where open circles represent the corresponding ionic positions after the displacement.

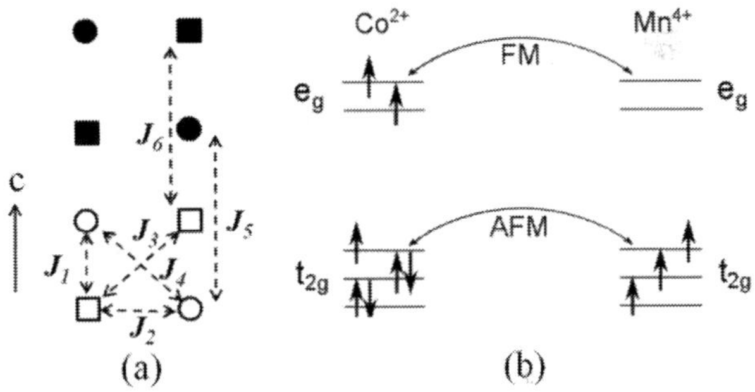

Figure 3. (a) The definition of exchange parameters in the ↑↑↓↓ magnetic order. (b)[23] Schematic crystal field level diagrams of the high-spin Co^{2+} and Mn^{4+}. The e_g electrons hopping leads to a FM coupling while the t_{2g} electrons hopping to a AFM coupling.

Unlike Y_2CoMnO_6, the ME behavior of Lu_2CoMnO_6 has reached consensus. The ME coupling in Lu_2CoMnO_6 has been reported by Yáñez-Vilar et al.[10] firstly, they report that Lu_2CoMnO_6 shows an E*-type magnetic order below 43 K, and ferroelectricity below 35 K that is strongly coupled to a net magnetism. However, their experiments are based on polycrystals, resulting in a question about whether ME behavior is intrinsic or is an extrinsic feature of polycrystal. Until recently, Chikara et al. have reported their results of electric polarization and magnetization measurements in Lu_2CoMnO_6 single crystals, which establish that the ME behavior of Lu_2CoMnO_6 is an intrinsic effect but not due to grain boundaries or other extrinsic effects of polycrystals[11]. Although the ME behavior of Lu_2CoMnO_6 is thus clarified, the direction of electric polarization is still controversial, as mentioned in Y_2CoMnO_6. The early experimental[10] results show that the polarization is along c-axis (Fig. 2(a)), parallel to the Co-Mn chains, while the newly experiment[11] reports that the electric polarization aligns along the b-axis (Fig. 2(b)), perpendicular to the magnetic field. Whether the electric polarization

Table 1. The reported exchange parameters J (meV) of Y_2CoMnO_6 and Lu_2CoMnO_6.

	J_1	J_2	J_3/J_4	J_5/J_6
Y_2CoMnO_6[23]	0.99	-4.5	0.14/0.22	
Y_2CoMnO_6[22]	0.678	-5.113	0.398/-0.878	-0.629/0.576
Lu_2CoMnO_6[24]	-2.8	-7.8	0.1	

along c or b axis will determine the ferroelectricity origin. The electric polarization along c is from the consideration of domain boundaries between ↑↑ and ↓↓ FM domains, producing an electric polarization within each domain boundary by exchange-striction induced inversion symmetry breaking(Fig. 2(a)). And the electric polarization along b is due to the O ions on the two neighboring chains shifting coherently along the b-axis (Fig. 2(b)) under the exchange striction. Based on the reported results from experiment and theory, it is more reasonable that the electric polarization of R_2CoMnO_6 should be along the b-axis.

As we mentioned above, both ferroelectricity origins are related to an exchange striction, which is certainly from the special ↑↑↓↓ magnetic order. Therefore, the origin of such ↑↑↓↓ order is the deep source of the ME property of R_2CoMnO_6. According to the ↑↑↓↓ origin of Ca_3CoMnO_6 that the NNN Co-Co/Mn-Mn exchange interactions larger than the NN Co-Mn plays an important role, the exchange interactions of R_2CoMnO_6 are widely investigated by theory. As shown in Table. I, the exchange interactions of NN Co-Mn intra-chain J_1, NN Co-Mn inter-chain J_2, NN Co-Co/Mn-Mn inter-chain J_3/J_4 and NNN Co-Co/Mn-Mn intra-chain J_5/J_6 (Fig. 3(a)) are calculated from the first-principles[23, 22, 24]. These values reflect that the NN Co-Mn inter-chain is a relatively strong FM interaction ($J_2 < 0$), which is significantly stronger than other interactions. Especially, the NN Co-Mn intra-chain interaction J_1 is a weak AFM exchange for Y_2CoMnO_6 and a weak FM exchange for Lu_2CoMnO_6, corresponding to the closest total-energy of FM and ↑↑↓↓ order. Such weak interactions could be disturbed by some other factors, thus resulting in the special ↑↑↓↓ order.

Then, the large magnetic exchange anisotropy and possible origin of ↑↑↓↓ order are discussed by analyzing the orbital occupations. From all the reported electronic structure of R_2CoMnO_6[23, 22, 24], the valences for Co and Mn ions are 4+ and 2+, respectively, and both are in high-spin state. The schematic energy level and possible interactions are illustrated in Fig. 3(b). The NN Co-Mn interactions are dominated by the FM superexchange (SE) via a hopping of Co^{2+} e_g electrons to the empty Mn^{4+} e_g states or half-occupied t_{2g} states. However, a AFM direct-exchange (DE) between the partially occupied Co^{2+} t_{2g} states and the half occupied Mn^{4+} t_{2g} states would weaken the FM exchange interaction. As we know, the SE is more influenced by the Co-O-Mn angle, while the DE is by the Co-Mn distance. Here, the Co-O-Mn angles along J_1 and J_2 are similar, but the Co-Mn distance along J_1 is more or less smaller than that along J_2. Therefore, the AFM DE is weaker than FM SE of the NN Co-Mn interchain (J_2) , where a robust FM interaction is formed. When the larger AFM DE is comparable with FM SE of the NN Co-Mn intrachain (J_1), the competition between the FM SE and AFM DE results in a weak AFM or FM interactions along Co-Mn chains. Based on the discussions above, our group have proposed that the ↑↑↓↓ ground state originates from the strong competition between the FM SE and AFM DE along J_1, like a frustration. As the temperature gets lower, the system chooses a crystal distortion or a Co-Mn dimerization to release the competition (frustration), because a definite AFM DE is dominant between the shortened Co-Mn sites while a FM interaction should be the case between the elongated Co-Mn sites[23]. Besides, Zhang et. al. have suggested that the Co-Mn antisite defects observed from experiment could introduce an extra Co-Co/Mn-Mn

AFM exchange, which may disturb the competition and result in the ↑↑↓↓ ground state[24].

3. R_2NiMnO_6

The structure of R_2NiMnO_6 is also monoclinic lattice, which is nearly the same with R_2CoMnO_6 (Fig. 1(c)). However, the magnetic ground state of R_2NiMnO_6 is an E-type configuration, which has been firstly predicted in Y_2NiMnO_6 from the first-principles design by Kumar et al.[15]. The E-type magnetic structure consists of ↑↑↓↓ spin chains in *ab* plane (Fig. 1(a)), with an out-of-plane FM coupling. Such E-type magnetism breaks inversion symmetry and generates a ferroelectric polarization. Meanwhile, the origin of the special E-type magnetic state is proposed by the competition between the AFM NNN Ni-Ni and FM NN Ni-Mn interactions. The frustration is released by increasing (decreasing) the distance between ↑-↑(↑-↓) along the ↑↑↓↓ spin chains and the E-type magnetic state is thus formed. This ME origin seems more reliable than that mentioned in R_2CoMnO_6, since the intraplane NNN Ni-Ni distance is much smaller than the interplane NNN Co-Co/Mn-Mn distance in R_2CoMnO_6. Thus, the NNN Ni-Ni interaction here could be comparable with the NN Ni-Mn interaction, which results in frustration and further E-type magnetism.

Nevertheless, such prediction has not been confirmed on consensus by experiment. The nanocrystal Y_2NiMnO_6 synthesized by sol-gel method exhibits a ferroelectric transition at 573 K, far above the Curie temperature of 92 K[25]. These results are not consistent with the theoretical prediction: for one hand, the ↑↑↓↓ magnetic order is not certain from experiment; for the other hand, the ferroelectricity is not driven by magnetostriction, where the ferroelectric transition temperature is far above the magnetic ordering temperature. Another experimental group has reported an appearance of ferroelectric polarization coinciding with the magnetic phase transition at about 67 K and has further proposed that the ferroelectricity originates from the Ni^{2+}/Mn^{4+} charge ordering and the ↑↑↓↓ spin ordering[26]. Such results indicate that the ferroelectricity is driven by E-type magnetism and is thus consistent with the theoretical prediction. However, a recent experiment has showed that Y_2NiMnO_6 displays a canted FM order at 81 K, not an E-type magnetism, and hence the ferroelectricity is absent[27]. Therefore, the magnetic ground state and related ME phenomenon of Y_2NiMnO_6 need more investigations to clarify.

Based on the consensus that the magnetic ground state should change from FM for La_2NiMnO_6 to AFM for Y_2NiMnO_6 due to a decrease of the R ionic size, most efforts are made to explore the double perovskites with small R ions. In_2NiMnO_6 with small In^{3+} ions (0.92Å) is thus prepared as a candidate for ME multiferroics[28, 29]. Yi et al. have reported that In_2NiMnO_6 is an AFM with a Neel temperature of 26 K from experiment, and have further indicated that the AFM order is E-type magnetic order by the first-principles calculations[28]. However, a noncollinear magnetic order has been obtained by Terada et al., which results in a ferroelectric polarization of $30\mu C/m^2$[29]. Accordingly, the magnetic ground state of In_2NiMnO_6 is also controversial.

4. Conclusions

All in all, the ↑↑↓↓ magnetism driven ferroelectricity is at the beginning. The exploration of new material is just on the way. More and more experiments are working to improve the quality of sample and the accuracy of numerical measure. Different theoretical methods are about to clarify the phenomenon and seek the deep origin. Although there are still many questions and controversy, we could predict that in the near future, this field will be very active both in basic physics and applications.

5. Acknowledgments
This work was supported by the National Science Foundation of China under Grant Nos. 11534012, and U1230202 (NSAF), the special Funds for Major State Basic Research Project of China (973) under Grant No. 2012CB933702 and Director Grants of CASHIPS. The calculations were partly performed in Center for Computational Science of CASHIPS, the ScGrid of Supercomputing Center and Computer Net work Information Center of Chinese Academy of Science, and partly using Tianhe-2JK computing time award at the Beijing Computational Science Research Center (CSRC).

References
[1] S.-W. Cheong and M. Mostovoy 2007 Nat. Mater. **6**, 13
[2] R. Ramesh and N. A. Spaldin 2007 Nat. Mater. **6**, 21
[3] D. Khomskii 2009 Physics **2**, 20
[4] W. Eerenstein, N. D. Mathur and J. F. Scott 2006 Nature (London) **442**, 759
[5] B. Lorenz, Y.-Q. Wang, and C.-W. Chu 2007 Phys. Rev. B **76**, 104405
[6] I. A. Sergienko, C. SŞen, and E. Dagotto 2006 Phys. Rev. Lett. **97**, 227204
[7] Y. J. Choi, H. T. Yi, S. Lee. Q. Huang, V. Kiryukhin, and S. W. Cheong 2008 Phys. Rev. Lett. **100**, 047601
[8] H. Wu, T. Burnus, Z. Hu, C. Martin, A. Maignan, J. C. Cezar, A. Tanaka, N. B. Brookes, D. I. Khomskii, and L. H. Tjeng 2009 Phys. Rev. Lett. **102**, 026404
[9] Y. Zhang, H. J. Xiang, and M.-H. Whangbo 2009 Phys. Rev. B **79**, 054432
[10] S. Yáñez-Vilar, E. D. Mun, V. S. Zapf, B. G. Ueland, J. S. Gardner, J. D. Thompson, J. Singleton, M. Sánchez-Andújar, J. Mira, N. Biskup, M. A. Señarís-Rodríguez, and C. D. Batista 2011 Phys. Rev. B **84**, 134427
[11] S. Chikara, J. Singleton, J. Bowlan, D. A. Yarotski, N. Lee, H. Y. Choi, Y. J. Choi, and V. S. Zapf 2016 Phys. Rev. B **93**, 180405(R)
[12] J. Blasco, J. L. García-Muñoz, J. García, J. Stankiewicz, G. Subías, C. Ritter, and J. A. Rodríguez-Velamazán 2015 Appl. Phys. Lett. **107**, 012902
[13] G. Sharma, J. Saha, S. D. Kaushik, V. Siruguri, and S. Patnaik 2013 Appl. Phys. Lett. **103**, 012903
[14] L. Wang, W. Zhou, D. Wang, Q. Cao, Q. Xu, and Y. Du 2015 J. Appl. Phys. **117**, 17D914
[15] S.Kumar, G. Giovannetti, J. van den Brink, and S. Picozzi 2010 Phys. Rev. B **82**, 134429
[16] W. Yi, Q. Liang, Y. Matsushita, M. Tanaka, and A. A. Belik 2013 Inorg. Chem. **52**, 14108
[17] I. O. Troyanchuk, D. D. Khalyavin, J.W. Lynn and R.W. Erwin, Q. Huang, H. Szymczak, R. Szymczak, and M. Baran 2000 J. Appl. Phys. **88**, 360
[18] M. Mouallem-Bahout, T. Roisnel, F. Bourée, G. André, C. Moure, and O. Peña 2007 Prog. Sol. State Chem. **35**, 257
[19] T. Chatterji, B. Frick and H. S. Nair 2012 J. Phys.: Condens. Matter **24**, 266005
[20] H. S. Nair, R. Pradheesh, Y. Xiao, D. Cherian, S. Elizabeth, T. Hansen, T. Chatterji, and Th. Bruckel 2014 J. Appl. Phys. **116**, 123907
[21] J. Blasco, J. García, G. Subías, J. Stankiewicz, J. A. Rodríguez-Velamazán, C. Ritter, J. L. García-Muñoz, and F. Fauth 2016 Phys. Rev. B **93**, 214401
[22] C. Y. Ma, S. Dong, P. X. Zhou, Z. Z. Du, M. F. Liu, H. M. Liu, Z. B. Yan, and J.-M. Liu 2015 Phys. Chem. Chem. Phys. **17**, 20961
[23] T. Jia, Z. Zeng, X. G. Li, and H. Q. Lin 2015 J. Appl. Phys. **117**, 17E119
[24] J. T. Zhang, X. M. Lu, X. Q. Yang, J. L. Wang, and J. S. Zhu 2016 Phys. Rev. B **93**, 075140
[25] R. P. Maiti, S. Dutta, M. Mukherjee, M. K. Mitra and D. Chakravorty 2012 J. Appl. Phys. **112**, 044311
[26] J. Su, Z. Yang, X. Lu, J. Zhang, L. Gu, C. Lu, Q. Li, J. Liu, and J. Zhu 2015 ACS Appl. Mater. Interfaces **7**, 13260
[27] H. Nhalil, H. S. Nair, C. M. N. Kumar, A. M. Strydom, and S. Elizabeth 2015 Phys. Rev. B **92**, 214426
[28] W. Yi, Q. Liang, Y. Matsushita, M. Tanaka, and A. A. Belik 2013 Inorg. Chem. **52**, 14108
[29] N. Terada, D. D. Khalyavin, P. Manuel, W. Yi, H. S. Suzuki, N. Tsujii, Y. Imanaka, and A. A. Belik 2015 Phys. Rev. B **91**, 104413

IOP Conf. Series: Journal of Physics: Conf. Series **827** (2017) 012006 doi:10.1088/1742-6596/827/1/012006

Magnetic phase diagram of FeO at high pressure

Peng Zhang[1], R. E. Cohen[2], K. Haule[3]

1. Department of applied Physics, Xi'An Jiaotong university, Xi'An, China
2. Carnegie Institution of Washington, Washington D.C. USA
3. Department of Physics, Rutgers university, Piscataway, New Jersey, USA

E-mail: zpantz@xjtu.edu.cn

Abstract. FeO is an insulator with anti-ferromagnetic (AFM) spin ordering at ambient pressure. At increased external pressure, the Néel temperature of FeO first increases at the pressure below 40 GPa. Experiments predict that the AFM ordering will collapse above 80 GPa, but the mechanism of the high pressure magnetic collapse is still unknown. Using the combination of density functional theory and dynamical mean-field theory (DFT+DMFT), the nature of the magnetic collapse of FeO is examined and its magnetic phase diagram up to 180 GPa is discussed.

1. Introduction

The electrical and magnetic properties of FeO at high pressure attracts long lasting attention. FeO is the endmember to magnesiowüstite, the second most abundant materials in Earth's mantle. FeO has cubic rock salt structure (B1) at ambient pressure [1]. As external pressure is increased, it first transferred to rhombohedrally distorted (rB1) structure at about 17 GPa and room temperature [2, 3], next to the B8 structure as pressure is further increased [4, 5, 6, 7]. At ambient condition, FeO is an insulator. At high pressure and temperature, it has been proved both theoretically and experimentally there is an insulator to metal transition [8]. At ambient pressure and below 200 K [9], the magnetic moments of iron in FeO ordered ferromagnetically in planes perpendicular to [111] direction, but ordered antiferromagnetically between adjacent planes. As pressure is increased, the Néel temperature T_N increases linearly at pressure below 40 GPa [10, 11]. At higher pressure (P>80 GPa), the magnetism of FeO collapse, whereas the mechanism of such collapse is in debating. Mössbauer spectroscopy experiment by Pasternak et al. suggests such magnetic collapse results from high spin to low spin transition (HS-LS) in iron atom and FeO is in LS state at P>80 GPa [12]. But later x-ray emission spectroscopy experiment by Badro et al. suggests the collapse of magnetism comes from antiferromagnetism to paramagnetism (AFM-PM) transition [13]. Badro et al. proposed an arch shape Néel temperature as a function of pressure, which peaks at around 50 GPa and the magnetic moments is finite up to 143 GPa. Such discrepancy calls for a theoretical proof. But the known numerical researches on the Néel temperature of FeO are mostly limited to the low pressure region [14]. Due to underestimation of strong correlations among d-electrons of iron, DFT fails to give out insulate FeO at ambient condition [15]. Here we employ $state-of-the-art$ density functional theory and dynamical mean-field theory (DFT+DMFT) [16, 17] in investigation of magnetic phase diagram of stoichiometric FeO in cubic B1 structure up to 200 GPa and 1000 K. We found two coexisted magnetic transitions in FeO when external pressure is increased, the on-site

IOP Conf. Series: Journal of Physics: Conf. Series **827** (2017) 012006 doi:10.1088/1742-6596/827/1/012006

HS to LS crossover in iron atom and the ordering (AFM) to non-ordering (PM) transition of iron atoms. The HS state of iron atom crossover to LS state due to the electron transformation from e_g orbitals to t_{2g} orbitals and the pairing of these d-electrons.

2. Methods

In DFT+DMFT, the crystal problem is treated by solving the DFT equations, whereas the strong correlations are included by solving the DMFT equations. The outputs of DFT with self-energy correction are used to construct impurity levels and hybridization function for the DMFT calculation. After the DMFT iteration is converged, new charge density and self-energy are used for next DFT iteration. Our DFT+DMFT formalism iterates until full convergence of charge, chemical potential and self-energy. The WIEN2K package is used for the DFT part. Hybridization expansion continuous time quantum Monte Carlo (CTQMC) method [18, 19] is used as impurity solver of the DMFT equation to introduce all local electron-electron scattering. More details for the DFT+DMFT method is referenced to [16, 17]. The parameterized Hubbard U of 3d electrons is 8.0 eV, and the corresponding Hund's J is calculated from the Yukawa screening.

In the paramagnetic region, the bulk magnetic susceptibility has Curie-Weiss behavior $\chi_q^m \propto \frac{1}{T-T_N}$ as the temperature is decreasing, and diverges at the Néel temperature $T = T_N$. In order to derive the bulk magnetic susceptibility from magnetic susceptibility on the DMFT impurity site, we calculated the magnetic vertex function from Bethe-Salpeter equation, which connects the bulk magnetic susceptibility with the impurity magnetic susceptibility.

$$1/\chi_{m,i}^0 - 1/\chi_{m,i} = \Gamma_m = 1/\chi_{m,q}^0 - 1/\chi_{m,q} \quad (1)$$

in which i is impurity index, q is moment, m indicates the magnetic susceptibility. Due to numerical efficiency, we truncated number of Fermionic frequency in our CTQMC measurements of the impurity magnetic susceptibility and in derivation of bulk magnetic susceptibility. We use bare magnetic susceptibility in real frequency to recover the lost part due to truncation of Fermionic frequency.

3. Results

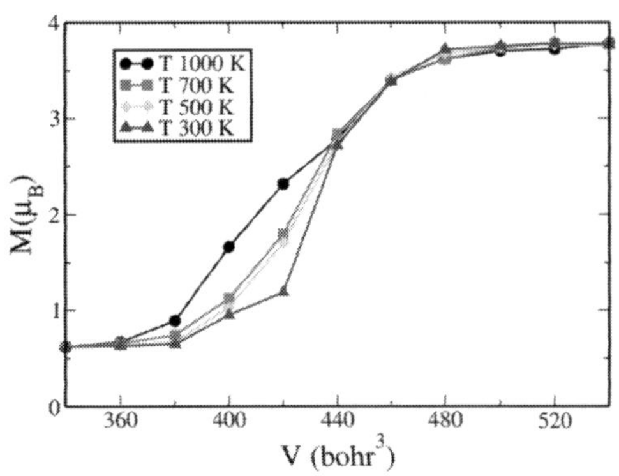

Figure 1: Magnetic moment of iron atom in FeO as a function of volume at temperature from 300 K to 1000 K. The magnetic moment of iron decreases as lattice volume is decreased, and the process of decrement is enhanced at lower temperature.

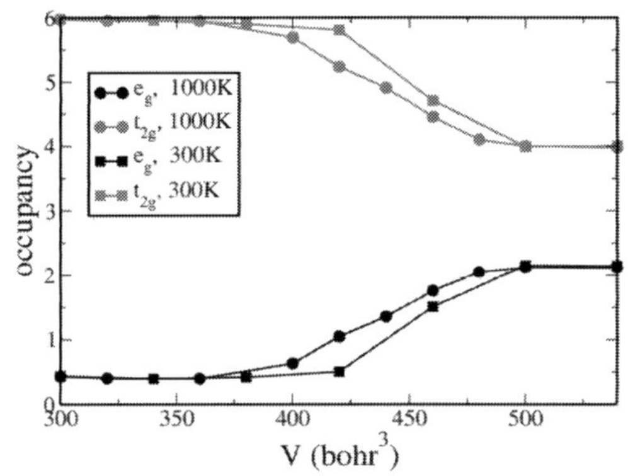

Figure 2: Electron occupancy on e_g and t_{2g} orbitals of iron atom in FeO at 300 K and 1000 K. As lattice volume is decreased, electrons on e_g orbitals transfer to t_{2g} orbitals. At lower temperature, the process of electron transfer is enhanced.

In Fig. 1, the magnetic moment $M = g\sqrt{\langle S_z^2\rangle}$ of iron in FeO is plotted as a function of lattice volume at temperatures from 300 K to 1000 K. As volume is compressed, magnetic moment of FeO gradually decreases. When temperature is decreased, the magnetic moment versus volume curves get steeper in the region of decrement. The magnetic moment versus volume curves converge at T<300 K. It indicates there is HS-LS transition in FeO as lattice volume is decreased, and the HS-LS transition is a crossover. The magnetic moments and the corresponding HS-LS transition in transition metal compounds usually result from partitioning and pairing of their d-electrons. This scenario is presented in Fig. 2. At V=540 $bohr^3$, there are about 4 electrons on t_{2g} orbitals and about 2 electrons on e_g orbitals. In atomic electron configuration, 4 up spins and 2 down spins give out atomic magnetic moment of $4\mu_B$. As volume is decreased from 500 $bohr^3$ to 360 $bohr^3$, electrons on e_g orbitals gradually transfer to t_{2g} orbitals. At 300 $bohr^3$, there is no electrons on e_g orbitals. The pairing of all 6 electrons on t_{2g} orbitals leads to the LS state. The speed of electron transfer is enhanced as temperature is decreased. The electron transfer from e_g orbitals to t_{2g} orbitals results from shifts of density of state (DOS) from e_g orbitals to t_{2g} orbitals, as shown in Fig. 3. In Fig. 3a, there are finite DOS on both e_g orbitals and t_{2g} orbitals above and below the Fermi level. The Fermi level falls in the gaps of e_g orbitals and t_{2g} orbitals, which indicates FeO is both Mott insulator and charge transfer insulator. The insulator state of FeO in our numerical calculation at the ambient condition is consistent with experiment results. In contrast, although in Fig. 3b FeO is still an insulator at 300 $bohr^3$, the DOS of e_g orbitals is zero below the Fermi level and the DOS of t_{2g} is zero above the Fermi level. There is no DOS available on e_g orbitals for electrons to occupy. All d-electrons are pushed to t_{2g} orbitals. FeO is no longer a Mott insulator at 300 $bohr^3$.

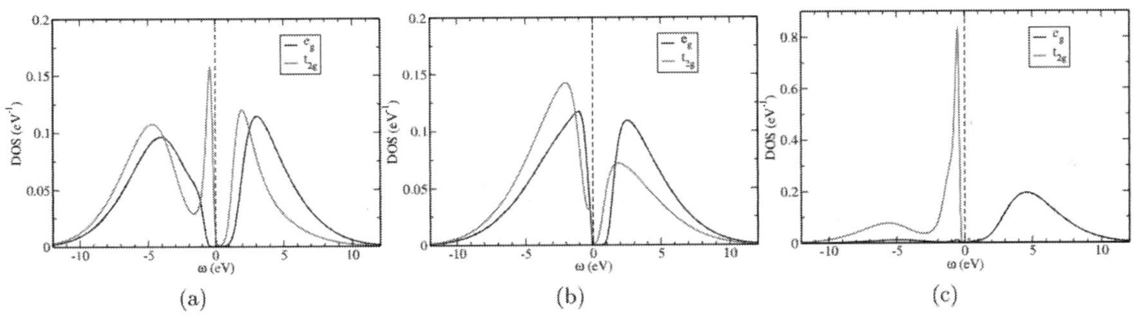

Figure 3: Density of states (DOS) on e_g and t_{2g} orbitals of iron atom in FeO at 300 K, (a) 540 $bohr^3$ (b) 420 $bohr^3$ (c) 300 $bohr^3$. As volume is decreased, DOS on e_g orbitals transfers to t_{2g} orbital, until at 300 $bohr^3$ there is no DOS below the Fermi level on e_g orbitals. FeO is always an insulator at all volumes.

Inverse bulk magnetic susceptibility χ_m^{-1} at $q = (\pi,\pi,\pi)$ as a function of temperature at various volumes in the paramagnetic region is plotted in Fig. 4. Since the bulk magnetic susceptibility $\chi_m(q=(\pi,\pi,\pi),T)$ behaves as $1/(T-T_N)$ in the paramagnetic region, which diverges at $T=T_N$, inverse of the bulk magnetic susceptibility $\chi_m^{-1}(q=(\pi,\pi,\pi),T)$ is linear with temperature. When we extrapolate the linear fit of $\chi_m^{-1}(q=(\pi,\pi,\pi),T)$ as a function of decreased temperature, the Néel temperature T_N is derived at the temperature where the linear fit crosses the temperature-axis. The derived Néel temperature at each volume is presented in legend of Fig. 4. The Néel temperature first increase as volume is compressed (or the pressure is increased); it peaks at 380 $bohr^3$ and then decrease at smaller volumes (or at higher pressures).

To summarize, we have find there are two magnetic transitions in FeO at increased pressure: a HS-LS transition as well as a AFM-PM transition. At above 80 GPa, the magnetic moment

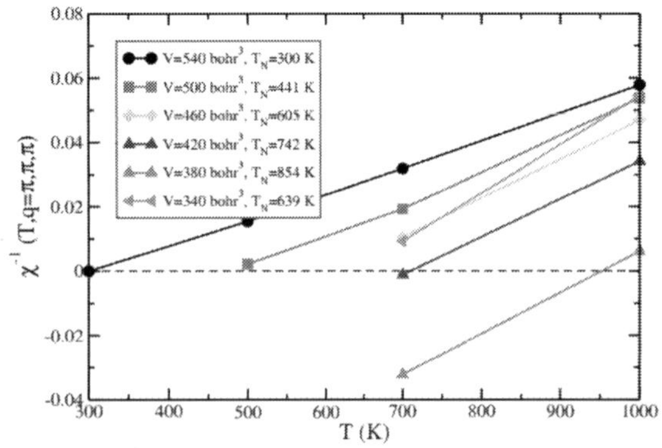

Figure 4: Inverse bulk magnetic susceptibility of FeO as a function of temperature at volumes from 540 $bohr^3$ to 300 $bohr^3$. The Nel temperature T_N is derived by linear extrapolation of $\chi^{-1}(T, q = (\pi, \pi, \pi))$ vs. T line to x-axis, where $T = T_N$. The corresponding Nel temperature T_N at each volume is presented in the legend.

of iron atom in FeO is nonzero. Our derived Néel temperature first increase then decrease with increased pressure. These results are consistent with previous experiment on FeO [20] and theoretical [21, 22] results of magnesiowstite.

3.1. Acknowledgments

This research used the NSF Extreme Science and Engineering Discovery Environment (XSEDE) supercomputer Stampede as well as the supercomputer Mw in Geophysical Laboratory, Carnegie Institution of Washington.

4. References

[1] Willis B T M and Rooksby H P 1953 *Acta Crystallogr.* **6** 827
[2] Zou G, Mao H K, Bell P M and Virgo D 1980 *Carnegie Inst. Washington Yearbook* **79**
[3] Yagi T, Suzuki T and Akimoto S 1985 *J. Geophys. Res.* **90** 8784
[4] Ozawa H, Hirose K, Tateno S, Sata N and Ohishi Y 2010 *Physics of the Earth and Planetary Interiors* **179** 157–163
[5] Fei Y and Mao H K 1994 *Science* **266** 1678
[6] Mazin I I, Fei Y, Downs R and Cohen R 1998 *Am. Mineral.* **83** 451
[7] Kondo T, Ohtani E, Hirao N, Yagi T and Kikegawa T 2004 *Phys. Earth Planet. Inter.* **201** 143–144
[8] Ohta K, Cohen R E, Hirose K, Haule K, Shimizu K and Ohishi Y 2012 *Phys. Rev. Lett.* **108** 026403
[9] McCammon C A 1992 *J. Magn. Magn. Mater.* **104** 1937–1938
[10] Nasu S, Kurimoto K, Nagatomo S, Endo S and Fujita F E 1986 *Hyperfine Interactions* **29** 1583
[11] Kantor I, Dubrovinsky L, Mccammon C, Dubrovinskaia N, Goncharenko I, Kantor A, Kuznetsov A and Crichton W 2007 *Phase Transitions* **80** 1151–1163
[12] Pasternak M P, Taylor R D, Jeanloz R, Li X, Nguyen J H and McCammon C A 1997 *Phys. Rev. Lett* **79** 5046
[13] Badro J, Struzhkin V V, Shu J, Hemley R J and Mao H 1999 *Phys. Rev. Lett.* **83** 4101
[14] Wan X, Yin Q and Savrosov S Y 2006 *Phys. Rev. Lett.* **97** 266403
[15] Cohen R E, Fei Y, Downs R, Mazin I I and Isaak D G 1998 **499** 27–40
[16] Kotliar G, Savrasov S Y, Haule K, Oudovenko V S, Parcollet O and Marianetti C A 2006 *Rev. Mod. Phys.* **78** 865
[17] Haule K, Yee C and Kim K 2010 *Phys. Rev. B* **81** 195107
[18] Werner P, Comanac A, de Medici L, Troyer M and Millis A J 2006 *Phys. Rev. Lett.* **97** 076405
[19] Haule K 2007 *Phys. Rev. B* **75** 155113
[20] Badro J, Fiquet G, Guyot F, Rueff J, Struzhkin V V, Vankó G and Monaco G 2003 *Science* **300** 789–791
[21] Tsuchiya T, Wentzcovitch R M, da Silva C R S and de Gironcoli S 2006 *Phys. Rev. Lett.* **96** 198501
[22] Holmström E and Stixrude L 2015 *Phys. Rev. Lett.* **114** 117202

IOP Conf. Series: Journal of Physics: Conf. Series **827** (2017) 012007 doi:10.1088/1742-6596/827/1/012007

Distance effect on magnetic coupling of Cr doped Bi_2Se_3: A first-principle and Monte Carlo study

Deyao Wu, Jingyao Liu, Ruqian Lian, Guigui Xu, Kehua Zhong, Jian-Min Zhang*, Zhigao Huang*

[1]Fujian Provincial Key Laboratory of Quantum Manipulation and New Energy Materials, College of Physics and Energy, Fujian Normal University, Fuzhou 350117, China

E-mail: jmzhang@fjnu.edu.cn; zghuang@fjnu.edu.cn

Abstract. In this paper, we investigated the magnetic properties of Cr doped Bi_2Se_3 combining Monte Carlo simulation with first principles method. Our results demonstrate that Cr atoms can be stably doped into Bi_2Se_3. Hysteresis loops were observed in the Cr doped system. The exchange interaction energies of magnetic atoms are calculated at the level of meV. The distance effect is studied with Cr atoms doped in different sites. The magnetic coupling of the atoms inner layer has great impact on magnetic materials, while the interlayer case has an obvious impact on coercivity of the material. However, as the distance of magnetic atoms exceed the quintuple layers (QLs), the magnetic coupling makes hardly contribution to the magnetic system. Our results may further promote the understanding of magnetic coupling in magnetic topological insulator.

1. Introduction

Topological insulator (TI) has become a research hotspot for its special quality that behaves as an insulator in its interior but whose surface contains conducting states [1-6]. Specifically, the prediction and preparation of three dimensions TIs make it of great interesting [7, 8]. Quantum anomalous Hall effect (QAHE) will occur when ferromagnetic order is introduced to this kind of materials [9-11]. In Cr doped Bi_2Se_3, the Curie temperature has an estimated value of 76 K at the doping level of 7.4% [11, 12]. Liu *et al.* observed an enhanced T_c of 50 K in $Bi_{2-x}Cr_xSe_3$ [13]. Ferromagnetism was observed in Cr doped Bi_2Se_3 by Chotorlishvili *et al.* [14], where the magnetic atoms prefer to orientate perpendicular to the TI surface. The antiferromagnetic order will appear by the magnetic impurity dopants of Gd in Bi_2Se_3 [15]. Zhao *et al.* found that Cr atoms adsorption on Bi_2Se_3 can induce a spin-polarization with total net magnetic moments of 2.157 μ_B [16]. Collins-McIntyre *et al.* reported a magnetic hysteresis loop with a coercive field of ~80 Oe and a ferromagnetic transition temperature of 20 K for 5.2% Cr doped Bi_2Se_3 [17]. However, Wang *et al.* showed that the Cr atoms deposition on Bi_2Se_3 surface would not lead to the gap opening at the Dirac point [18]. On the other hand, the electrical field tunning of magnetism was greatly researched in the magnetic topological insulators. Wang *et al.* proposed that electric field could induce the quantum phase transition from ferromagnetism to paramagnetism [19]. Wei *et al.* focused on the $Bi_{2-x}Mn_xSe_3$ crystals topological insulator and found that the concentration of Mn showed metal behavior when $x<0.05$ and it became *p*-type semiconductor at $x=0.07$ [20].

Actually, the observing QAHE still have many problems which have not been solved. For example, the topologically nontrivial band gap opened is too narrow, QAHE can only be found under ~ mK. In addition, the mechanism of QAHE was still controversial. To figure out these problems, we have studied the magnetic coupling of Cr atoms in Bi_2Se_3 using first-principle calculations and Monte Carlo simulation.

2. Method and Model

2.1. Method

Generally, the magnetic states in magnetic materials can be determined by several competing energies: exchange energy, dipolar interaction, anisotropy energy, Zeeman energy. Therefore, the Hamiltonian of Cr-doped topological insulators can be described by

$$E = -\sum_{ij} J_n \overline{S}_i \overline{S}_j + D[\frac{\overrightarrow{S_i} \cdot \overrightarrow{S_j}}{r_{ik}^5} - 3\frac{(\overrightarrow{S_i} \cdot \overrightarrow{r_{ik}})(\overrightarrow{S_j} \cdot \overrightarrow{r_{ik}})}{r_{ik}^5}] - K\sum_i (\overline{S}_i \cdot \overrightarrow{u_i})^2 - H\sum_i S_i^z \qquad (1)$$

Where J_n is the exchange coupling constant, J_1 is the inner layer exchange constant and J_2 is the interlayer layer exchange constant, while J_3-J_7 is the different QLs exchange constant; D is dipolar interaction constant, here $D=0$; K is the anisotropy constant [21, 22]; H is the magnetic field intensity. The magnetization per atom is calculated by $M =< [(\sum_i S_i^x)^2 + (\sum_i S_i^y)^2 + (\sum_i S_i^z)^2]^{1/2} > /N$. We use Monte Carlo (*MC*) to analyze the magnetization reversal processes for Cr doping, where the spins can rotate freely in three dimensions.

2.2. Model

As shown in Figure 1, it is the crystal structure for 3×3×1 Bi_2Se_3 supercell in our VASP calculation. Two Cr atoms are replaced for different Bi atoms and the doping ratio is 3.703%. Projected augment wave (PAW) potentials with Perdew-Burke-Ernzerh (PBE) type of gradient approximation (GGA) is used in the first-principles calculations. Spin-orbital coupling (SOC) is considered. The lattice constants a=4.138 Å and c=28.64 Å for Bi_2Se_3. The cutoff energy was set at 300 eV. The Brillouin zone for the 3×3×1 Bi_2Se_3 is sampled by a gamma-centered 7×7×2 mesh.

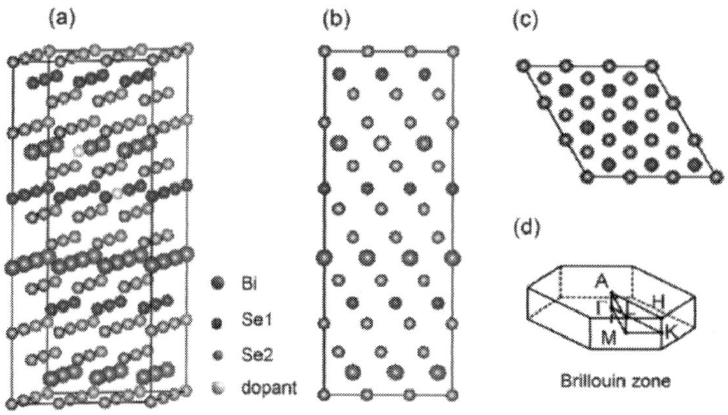

Figure 1. (a) Crystal structure illustration of a 3×3×1 supercell for the modelling of two dopants in bulk Bi_2Se_3 with (b) the side view and (c) the top view. (d) Brillouin zone and high symmetry points of the 3×3×1 supercell.

3. Results and Discussions

3.1. Electronic property of Cr doped Bi_2Se_3

We calculated band structure for Cr doped Bi_2Se_3 with the concentration of 4%, as show in Figure 2. We find an inverted band at the G point due to the strong SOC effect in this material, the opened gap for Cr doped Bi_2Se_3 is 0.08 eV and the magnetic moment is about 3 μ_B. The result shows the magnetic

IOP Conf. Series: Journal of Physics: Conf. Series **827** (2017) 012007 doi:10.1088/1742-6596/827/1/012007

atoms of Cr can be effectively doped in Bi_2Se_3, ferromagnetic order will appear and the material is still insulator, rendering it to be an ideal material to realize QAHE.

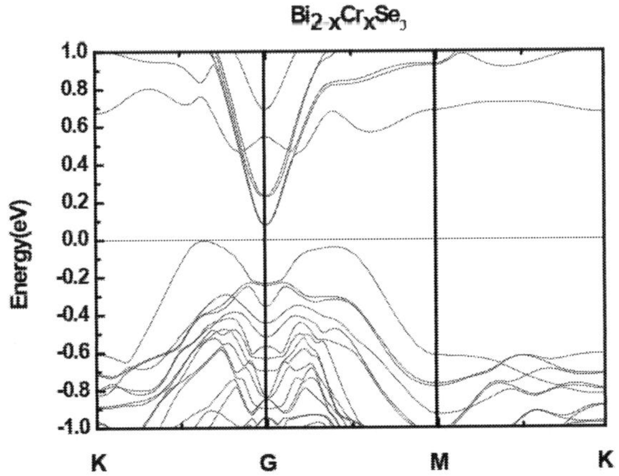

Figure 2. SOC band structure for relaxed Cr doped Bi_2Se_3 concentration of 4% ($x = 0.083$).

3.2. The magnetic coupling of the atoms inner layer

To investigate the magnetic coupling of doped magnetic atoms, two Cr atoms are substituted for Bi atoms, as seen in Figure 1. Our first-principle calculation shows the nearest magnetic coupling strengths is 22 meV, while the strengths after J_7 are less than 2 meV [12]. Therefore, we mainly focus on the distance effect on the magnetic coupling in $Bi_{2-x}Cr_xSe_3$.

As shown in Figure 3, we can find that when the inner layer exchange constant J_I changes from -1 to 0, the saturation magnetization gradually increased. As inner layer exchange constant J_I close to zero, the magnetic hysteresis loop is easier saturated. In contrast, when J_I varies from 0 to 1, the magnetization gradually decreased. What's more, the hysteresis loop arise obviously hysteresis phenomena, and the residual magnetism (M_r) and coercivity (H_c) are not zero. This is due to that system will show antiferromagnetic property when J_I is negative, accompanying with M_r and H_c close to 0. On the contrary, when J_I is positive, the material exhibits ferromagnetism with the obviously hysteresis effect emerging. The saturation magnetization (M_s) comes up with a maximum when J_I=0.1.

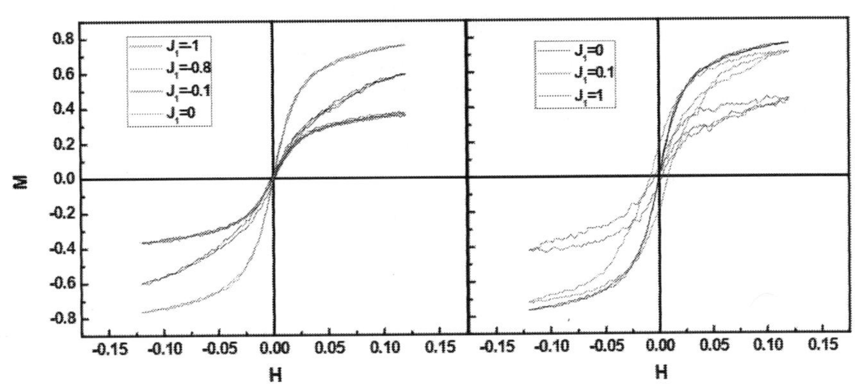

Figure 3. Hysteresis loops of Cr doping Bi_2Se_3 with different J_1 .

3.3. The magnetic coupling of the atoms interlayer layer

From Figure 4, we can see the effect of J_2 is same to J_1. While, M_r and H_c will tend to be zero as J_2 changes from 0 up to 0.2. In addition, M_s, M_r and H_c almost come up to the maximum value at the same time when J_2=0.1, as can be seen in Figure 5.

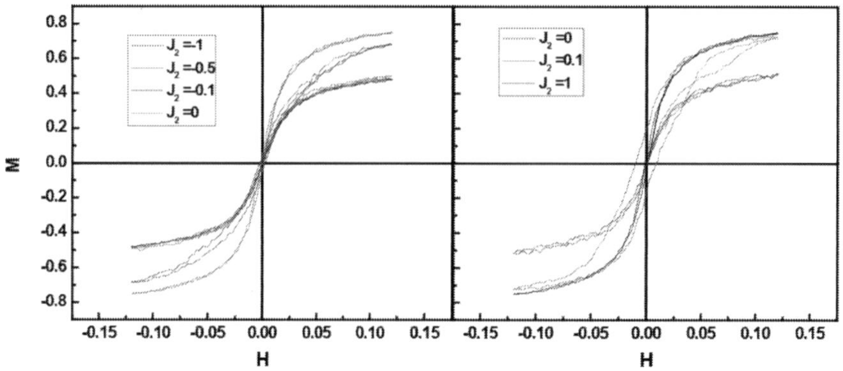

Figure 4. Hysteresis loops of Cr doping Bi_2Se_3 with different J_2.

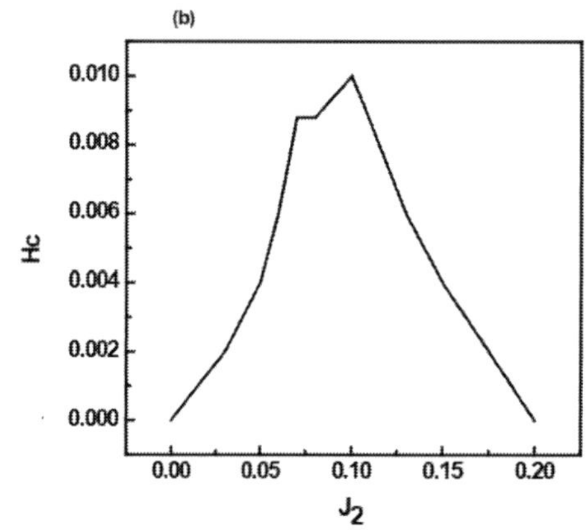

Figure 5. The variation of H_c with different J_2.

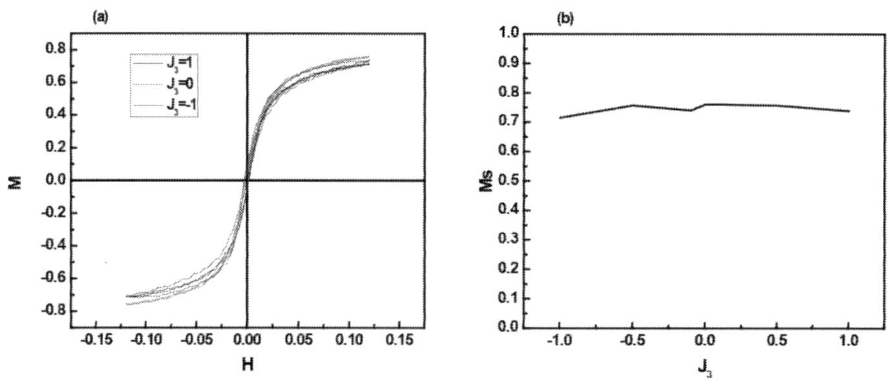

Figure 6. (a) Hysteresis loops of Cr doping Bi_2Se_3 with different J_3, (b) the variation of saturation magnetization M_s with different J_3.

3.4. The magnetic coupling of the atoms in different QLs

Figure 6(a) shows hysteresis loops hardly change when two Cr atoms are doped in different QLs. Therefore, the magnetic coupling between different QLs has little contribution to the system. It is also confirmed in Figure 6(b), the values of M_s and M_r are always at about 0.71 and 0.08, respectively.

Coercivity H_c is always zero. The reason is that the exchange energy is too little as the doping atoms far apart. This result proves that J_3-J_7 have little affection to the system, which is also confirmed by our first-principle calculations [12].

4. Conclusion

In summary, we have studied the distance effect on the magnetic coupling for Cr doped Bi_2Se_3 using first-principles method and Monte Carlo simulation. We found that the hysteresis loops is sensitive to the direct exchange energy between magnetic atoms. When two Cr atoms exist within the same QL, the system shows to be antiferromagnetism when J_1 and J_2 are negative, while ferromagnetism is appear when J_1 and J_2 are positive. However, as the magnetic atoms distribute exceed 1QL, the exchange energies have little contribution to the magnetism. Our results provide important guidelines towards the further efforts of magnetic doping in TIs, in particular, for the realization of QAHE.

Acknowledgments

We acknowledge the financial support by the National Natural Science Foundations of China (Nos. 61574037, 21203025, 11344008, 11404058, 61404029) and the Natural Science Foundations of Fujian Province of China (Nos. 2015J05005, 2016J01011, 2016J05151)

Reference

[1] Hasan M Z and Kane C L 2010 *Rev. Mod. Phys.* **82** 3045–3067; Qi X L and Zhang S C 2011 *Rev. Mod. Phys.* **83** 1057
[2] Kane C L, Mele E J 2005 *Phys. Rev. Lett.* **95** 146802
[3] Bernevig B A, Hughes T L, Zhang S C 2006 *Science* **314** 1757–1761
[4] Moore J E, Balents L 2007 *Phys. Rev. B: Condens. Matter* **75** 121306
[5] Liu C X, Qi X L, Dai X, Fang Z, Zhang S C 2008 *Phys. Rev. Lett.* **101** 146802
[6] Tse W K, MacDonald A H 2010 *Phys. Rev. Lett.* **105** 057401
[7] Zhang H, Liu C X, Qi X L, Dai X, Fang Z and Zhang S C 2009 *Nat. Phys.* **5** 438
[8] Xia Y, Qian D, Hsieh D, Wray L, Pal A, Lin H 2009 *Nat. Phys.* **5** 398
[9] Yu R, Zhang W, Zhang H J, Zhang S C, Dai X and Fang Z 2010 *Science* **329** 61
[10] Chang C Z *et al.* 2013 *Science* **340** 167
[11] Zhang J M, Zhu W, Zhang Y, Xiao D and Yao Y 2012 *Phys. Rev. Lett.* **109** 266405
[12] Zhang J M *et al.*2013 *Phys. Rev. B* **88** 235131
[13] Liu W. He L, Xu Y B *et al.* 2015 *Nano Lett.* **15** 764–769
[14] Chotorlishvili L, Ernst A, Dugaev V K 2014 *Phys. Rev. B* **89** 075103
[15] Kim S W, Vrtnik S, Dolinsek J and Jung M H 2015 *Appl. Phys. Lett.* **106** 252401
[16] Zhao X *et al.* 2014 *Physica E* **55** 9–12
[17] Collins-McIntyre L J *at al.* 2014 *EPL* **107** 57009
[18] Wang E, Tang P and Wan G L 2015 *Nano Lett.* **15** 2031–2036
[19] Wang J, Lian B and Zhang S C 2015 *Phys. Rev. Lett.* **115** 036805
[20] Wei Z T, Lv L, Zhang M, Yang X S, Zhao Y 2015 *J Supercond Nov Magn* **28** 2083–2088
[21] Binder K *et al.* 2010 *Monte Carlo Simulation in StatisticalPhysics: An Introduction, 5th ed*
[22] Wu Q *et al.* 2007 *Solid State Commun* **142** 242

IOP Conf. Series: Journal of Physics: Conf. Series **827** (2017) 012008 doi:10.1088/1742-6596/827/1/012008

The local magnetic moment and electron transfer of ZnO-based dilute magnetic semiconductors

Bin Zhuang, Yanmin Yang, Kehua Zhong, Jian-Min Zhang, Guigui Xu, Zhigao Huang*

[1]Fujian Provincial Key Laboratory of Quantum Manipulation and New Energy Materials, College of Physics and Energy, Fujian Normal University, Fuzhou 350117, China

E-mail: zghuang@fjnu.edu.cn

Abstract. The electronic structures and magnetic properties of ZnO semiconductors doped with Cu, Co, C, Al and S are studied by first-principles calculation. The electronic transfer among Zn, O and doped atoms, and the differences of the number of electron between spin-up and spin-down, △s, △p, △d for *s*, *p* and d orbits of these atoms, are analyzed in detail. It is found that, the ferromagnetic ground state is stabilized by its half-metallic electronic structure, and the strong local magnetic moments in $Zn_{1-x}Co_xO$, $Zn_{1-x}Cu_xO$ and $ZnO_{1-x}C_x$ (x = 5.55%) DMSs originate mainly from the strong hybridizations between Cu-3d and O-2p, Co-3d and O-2p, Zn-3d and C-2p electrons. It is considered that the requirements to give rise to the ferromagnetism in the DMSs are the strong local magnetic moment and the electron transfer. The magnetic coupling in $Zn_{1-x}Co_xO$, $Zn_{1-x}Cu_xO$ and $ZnO_{1-x}C_x$ is also considered to be a RKKY interaction.

1. Introduction

Up to now, room-temperature ferromagnetism (FM) has been observed in ZnO doped with V [1], Cr [2], Mn [3], Fe [4], Co [5-6], Ni [7], Cu [8] and C[9]. There are various viewpoints about the mechanism of ferromagnetism in DMSs. First viewpoint considers that the ferromagnetism comes from the secondary phase [10]. The second one suggested that the high-temperature ferromagnetism originates from O-vacancy-stabilized metastable phase rather than through a carrier-induced mechanism between separated and uniformly dispersed Mn atoms in ZnO [11]. The third one is that, the ferromagnetic state is stable due to the double-exchange mechanisms [12]. The fourth one is that, the ferromagnetic state is stable due to the long-range coupling mechanisms [13-15]. Priour et al. considered ferromagnetism in spatially randomly located magnetic moments, as in a diluted magnetic semiconductor, coupled via the carrier-mediated indirect exchange RKKY interaction [13]. The calculated results on the O surfaces of Co-ZnO indicated that substitutional Co ions develop large magnetic moments, and its long-range order depends on their mutual distance [14]. Zhu et al.suggested that Li and Cu atoms tend to segregate toward Mn atoms and strongly promote the ferromagnetic coupling via either RKKY or superexchange interaction [15].

However, up to now, the ferromagnetic coupling mechanism of DMSs is still not quite clear. Is it short range exchange, super-exchange coupling or long-range RKKY exchange? In this paper, the electronic structures and the magnetic properties of ZnO semiconductors doped with Cu, Co, C, Al and S are studied by first-principles calculation.

2. Models and calculation

The electronic structures and magnetic coupling in $Zn_{1-x}M_xO$ or $ZnO_{1-x}M_x$ (x=5.55%, M=Co, Cu, C, Al, S) were calculated based on DFT. The first-principles calculations were performed using PAW method as implemented in VASP. The detailed calculations can be also found in our previous papers [16,17]. A plane-wave energy cutoff of 520 eV was used. The relaxations of lattices were stopped until the forces on each ion are converged to less than 0.01 eV/Å. These parameters ensure a good convergence of 1 meV on the total energies. The experimental lattice constants of ZnO unit cell (a = b = 3.249 Å, c = 5.205 Å) were used. To study the electron transfer, the positions of 72 atoms are labeled in Figure 1 for x=5.55%. Here, both O atoms at the 8th and ith (i=25, 14, 33, 7, 35, 18, 9) sites are replaced by two C or S (grey) atoms; It gives seven distinct pairs of M atoms, such as (8, 25), (8, 14), (8, 33), (8, 7), (8, 35), (8, 18), (8, 9), respectively. Both Zn atoms at the 46th and ith (i=57, 38, 55, 45, 61, 50, 48) sites are substituted with two Co (Cu or Al) atoms. It gives seven distinct pairs of M atoms, such as (46, 57), (46, 38), (46, 55), (46, 45), (46, 61), (46, 50), (46, 48), respectively.

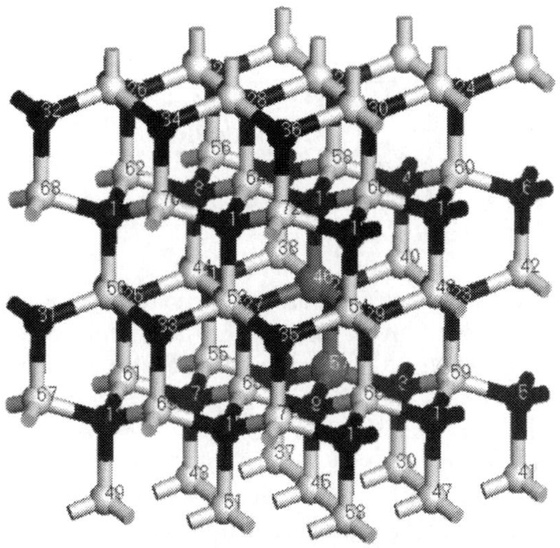

Figure 1. The 3 × 3 × 2 supercell of wurtzite ZnO containing 36 Zn (blue) and 36 O (red) atoms with the labeled doped atomic sites.

3. Results and discussions

The typical total density of states (DOSs) and projected density of states (PDOSs) for $Zn_{0.9445}Cu_{0.0555}O$, have been calculated. From the calculated results, it is found that the their electronic structures have a half-metallic character, and most of the spin polarization states are found to originate from Cu 3d electrons. Also, the majority of the peaks at the Fermi level is contributed by Cu-3d electrons. Moreover, the ferromagnetism is a consequence of overlap between Cu 3d and O 2p states in the vicinity of the Fermi level, namely hybridization between Cu 3d electron and O 2p electron. For $Zn_{0.9445}Co_{0.0555}O$ and $ZnO_{0.9445}C_{0.0555}$, the similar DOSs, PDOSs and ferromagnetism have been found in our previous studies [5, 17]. The total magnetic moments for the $Zn_{1-x}Co_xO$, $Zn_{1-x}Cu_xO$ and $ZnO_{0.9445}C_{0.0555}$ supercells are nearly 6.05μ_B, 2.0μ_B and 4.0μ_B respectively. The results above are coincident with Akai's theory in which it is suggested that DMS with half-metallic character has a ferromagnetic ground state [12]. So the nearly half-metallic electronic structures should favor the ferromagnetic ground state. The typical DOSs and PDOSs for $Zn_{0.9445}Al_{0.0555}O$ and $ZnO_{0.9445}S_{0.0555}$ have also been calculated. The calculated results indicate that the DOSs and PDOSs of spin-up and spin-down are almost symmetrical, which means that ZnO DMSs doped with Al or S are not magnetic.

For each configuration in Figure 1, the total energy difference $J= E_{AFM}-E_{FM}$ between the ferromagnetic (E_{FM}) and antiferromagnetic (E_{AFM}) spin configurations for various doped concentrations are calculated. In the simulation, the value of J is considered as the magnetic coupling strength for the pair of Co, Cu or C atoms. If the value of J is negative, the AFM configuration is more stable, and if it is positive, the FM configuration is more stable. Figure 2 shows the magnetic coupling strength *J* of $Zn_{0.945}Co_{0.055}O$, $Zn_{0.945}Cu_{0.055}O$ and $ZnO_{0.945}C_{0.055}$ for seven Co-Co, Cu-Cu and C-C configurations. From the figure, it is found that most of the configurations of $Zn_{0.945}Co_{0.055}O$, $Zn_{0.945}Cu_{0.055}O$ and $ZnO_{0.945}C_{0.055}$ DMSs have FM ground state except configuration 1 of $ZnO_{0.945}C_{0.055}$. This trend is consistent with the results obtained by Sluiter et al. [18] and Park et al.[19].

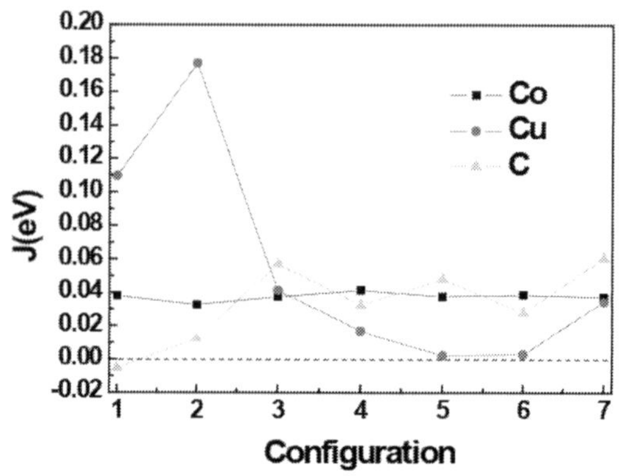

Figure 2. Magnetic coupling strength *J* for seven Co-Co, Cu-Cu and C-C configurations for $Zn_{0.945}Co_{0.055}O$, $Zn_{0.945}Cu_{0.055}O$ and $ZnO_{0.945}C_{0.055}$.

Figures. 3 (a) (b) (c) show the electron transfer structures of configurations 2, 5 and 7 for $Zn_{0.945}Cu_{0.055}O$, and pure ZnO. As seen in Figure 2, the configurations 2, 5 and 7 have all stable ferromagnetism. For the pure ZnO, every O atom gains 1.25 electrons, and every Zn atom loses 1.25 electrons. When Cu atoms are doped, the electron transfer resulting from the doped Cu atoms is evident, and the charge redistribution appears. The electrons of doped local Cu atom are transfered not only to the nearest O atoms but also to O atoms being for away the local Cu atom, which means that the transfer electrons above are itinerant. Similar electron transfer structures have been obtained in $Zn_{0.945}Co_{0.055}O$ and $ZnO_{0.945}C_{0.055}$. Comparing to the electron transfer structures of three samples, we can find that the itinerant electron number in $Zn_{0.945}Co_{0.055}O$ is minimum, and one in $Zn_{0.945}Cu_{0.055}O$ is second, and one in $ZnO_{0.945}C_{0.055}$ is largest.

In order to investigate further the exchange coupling mechanism in DMSs, we choose the atom plane containing first doped atom, which is named as "**both atom layers**". Figures 4(a) and (b) show the typical space distributions of the atoms in the both atom layers above for the configuration 1 of $ZnO_{0.945}C_{0.055}$, and the configuration 2 of $Zn_{0.945}Co_{0.055}O$ (or $Zn_{0.945}Cu_{0.055}O$), respectively. Moreover, the difference of spin-up and spin-down electrons in *s*, *p*, and *d* orbits of those Zn,O and Co(or Cu, C) in both atom layers above have been calculated. Figure 5 (a) shows the differences of spin-up and spin-down electrons, $\triangle s$, $\triangle p$, $\triangle d$ in *s*, *p*, and *d* orbits of Co, Zn and O in the "**both atom layers**" of configuration 4 for $Zn_{0.945}Co_{0.055}O$. Figure 6 (b) shows ones of Cu, Zn and O in the "**both atom layers**" of configuration 2 for $Zn_{0.945}Cu_{0.055}O$. Figure 5 (c) shows ones of C, Zn and O in the "**both atom layers**" of configuration 7 for $ZnO_{0.945}C_{0.055}$. From Figure 5(a), it is found that the local magnetic moment of the doped Co atoms in $Zn_{0.945}Co_{0.055}O$ is largest, and the local doped Co atoms induce the strong 3d moment of Zn and 2s (2p) moments of O, which gives rise to the largest total moment with $6.05\mu_B$. From Figure 5(c), one notices that, although the local 2p moment of the doped C atoms in $ZnO_{0.945}C_{0.055}$ is minimum, the doped C atoms induce the considerable 3d moment of Zn and 2s moments of O, which results in the secondary total moment with $4.00\mu_B$. From Figure 5(b), we can observe that, the local magnetic moment of the doped Cu atoms in $Zn_{0.945}Cu_{0.055}O$ is moderate, but the doped Cu atoms induce the least 3d moment of Zn and 2s moments of O, which leads to the minimum total moment with $2.00\mu_B$.

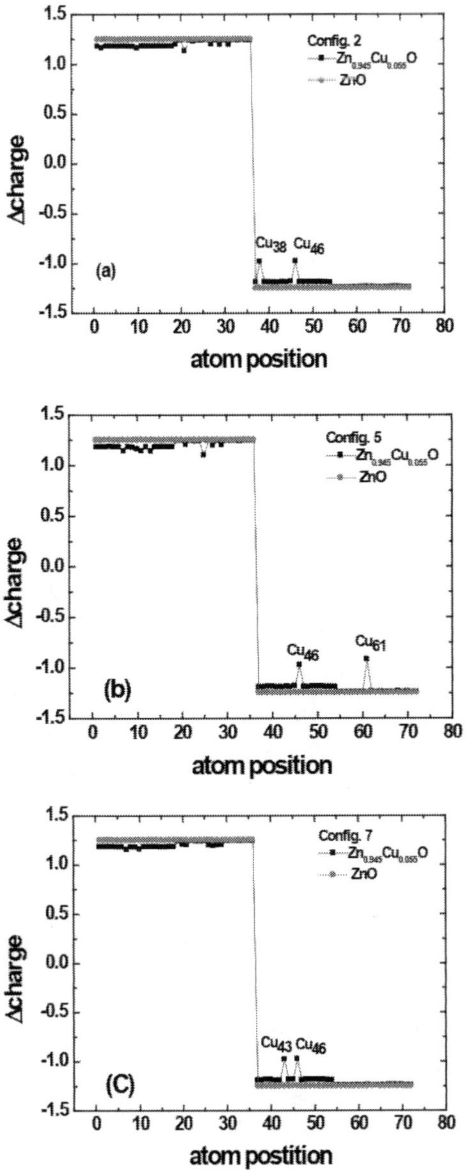

Figure 3. The electron transfer strctures of configurations 2, 5 and 7 for $Zn_{0.945}Cu_{0.055}O$ DMSs, and pure ZnO.

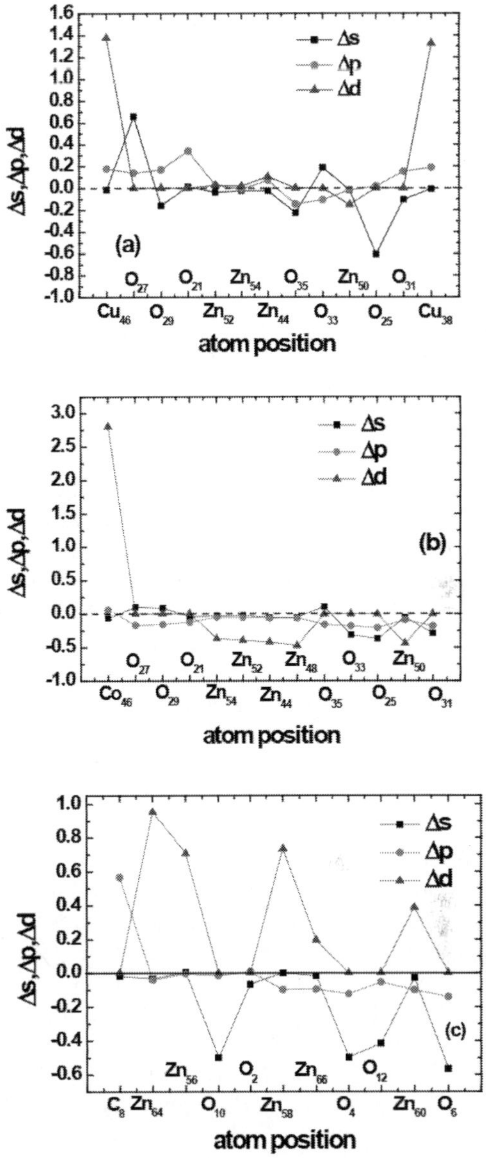

Figure 5. The differences of spin-up and spin-down electrons in s, p, and d orbit of Co, Zn and O for **"both atom layers"**: (a) confi. 4 of $Zn_{0.945}Co_{0.055}O$; (b) confi. 2 of $Zn_{0.945}Cu_{0.055}O$; (c) confi. 7 of $ZnO_{0.945}\ C_{0.055}$.

The electron transfer structures for $Zn_{0.945}Al_{0.055}O$ and $ZnO_{0.945}S_{0.055}$ indicate that, after two O atoms are replaced by two Al and S atoms in ZnO, the charge distributions of Zn and O atoms have the evident change, which means that the doped Al or S atoms give rise to the obvious electron transfer. However, it is observed that the values of $\triangle$s, $\triangle$p, $\triangle$d are almost near zero, which means that there does hardly exist the localized magnetic moment in $Zn_{0.945}Al_{0.055}O$ and $ZnO_{0.945}S_{0.055}$.

From the discussions above, it is found that the requirements to give rise to the ferromagnetism in the DMSs are the strong local magnetic moment and the electron transfer between the doped atoms,

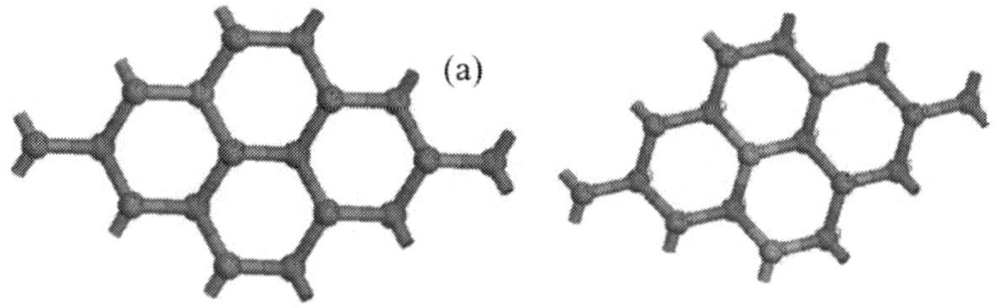

Figure 4. Atom array structure of the both atom layers (a) for the confi. 1 of $ZnO_{0.945}C_{0.055}$ and (b) for the confi. 2 of $Zn_{0.945}Co_{0.055}O$ (or $Zn_{0.945}Cu_{0.055}O$).

such as Co, Cu and C. The locale moments result from the strong hybridization between the Cu-3d and O-2p electrons, Co-3d and O-2p electrons, Zn-3d and C-2p electrons, respectively. Moreover, in $Zn_{1-x}(Co,Cu)_xO$ and $ZnO_{1-x}C_x$, there exists the electron transfer ont only among the doped atoms and the nearest neighbor atoms, but also among the doped atoms and the long-range atoms. That is, the magnetic coupling among the doped atoms (the local magnetic moments) is implemented by the interaction between the local magnetic moments and the transferred itinerant electrons. Therefore, the *p-d* hybridization and the itinerant d carriers are suggested to mediate the long range ferromagnetic exchange between Co, Cu and C atoms. The magnetic coupling in $Zn_{1-x}(Co,Cu)_xO$ and $ZnO_{1-x}C_x$ is considered to be a RKKY interaction. Although there exists the obvious electron transfer in $Zn_{1-x}Al_xO$ and $ZnO_{1-x}S_x$, there is not local moment. As a result, $Zn_{1-x}Al_xO$ and $ZnO_{1-x}S_x$ have not ferromagnetism.

4. Conclusion

The electronic structures and magnetic properties of $Zn_{1-x}M_xO$ (M=Co, Cu, Al, C, S) (x = 5.55%) DMSs have been studied. The requirements to give rise to the ferromagnetism in the DMSs are the strong local magnetic moment and the electron transfer between the doped atoms. The magnetic coupling in $Zn_{1-x}Co_xO$, $Zn_{1-x}Cu_xO$ and $ZnO_{1-x}C_x$ is considered to be RKKY interaction.

References

[1] Xiong Ze et al., 2011 Appl. Phys. Lett. 99, 052513
[2] Weng Z Z et al., 2012 J. Appl. Phys. 111, 113915
[3] Xu H Y et al., 2006 J. Chem. Phys. 124, 074707
[4] Shiv M S et al., 2011 J. Appl. Phys. 110, 103508
[5] Wu Q Y et al., 2007 Solid State Communications, 142(4), 242
[6] Qi S F et al., 2011 Phys. Rev. B 84, 205204
[7] Zhou S Q et al., 2006 J. Appl. Phys. 100, 114304
[8] Khan Z A et al., 2011 Appl. Phys. Lett. 99, 042504
[9] Pan H et al., 2007 Phys.Rev.Lett. 99.127201
[10] Shim J H et al., 2005 Appl. Phys. Lett. , 86, 082503
[11] Kundaliya D C et al., 2004 Nature Materials ,709
[12] Akai H, 1998 Phys. Rev. Lett , 81, 3002
[13] Priour D J, Jr, Das Sarma S, 2006 Phys. Rev. Lett. 97, 127201
[14] Sanchez N, Gallego S, Munoz M C, 2008 Phys. Rev. Lett. 101, 067206
[15] Zhu Y, Cao J X, Yang Z Q, Wu R Q, 2009 Phys. Rev. B 79, 085206
[16] Xu G G et al., 2008 Phys. Rev. B 78, 115420
[17] Chen San et al., Acta Phys. Sin. 58(2009)2011(in Chinese)
[18] Park M, Min B I, 2003 Phys.Rev.B 68, 224436
[19] Sluiter M H F et al., 2005 Phys. Rev. Lett. 94, 187204

Acknowledgments: We acknowledge the financial support by the National Natural Science Foundations of China (Nos. 61574037, 21203025, 11344008, 11204038, 11404058, 61404029)

Phase diagrams of the Kane-Mele-Hubbard model in the presence of an external magnetic field

Ya-Min Quan[1],Wei-Cheng Bao[1],Wei-Hua Wang[2] and Da-Yong Liu[1]

[1] *Key Laboratory of Materials Physics, Institute of Solid State Physics, Chinese Academy of Sciences, Hefei 230031, China*

[2] *Department of Electronics and Tianjin Key Laboratory of Photo-Electronic Thin Film Devices and Technology, Nankai University, Tianjin 300071, China*

E-mail: ymquan@theory.issp.ac.cn

Abstract. We studied the Kane-Mele-Hubbard model at half filling with Zeeman magnetic field using rotationally invariant slave boson method. The competition between external magnetic field and spin orbital coupling at large Hubbard U is explored. It is found that different canted antiferromagnetic phases exist in the presence of longitudinal and transverse magnetic fields. With longitudinal external magnetic field and strong electron correlation, the z component of the local moment is preferred by the finite spin-orbital coupling.

1. Introduction

The transition metal oxides such as Iridates have been investigated extensively due to the various physical phenomena driven by the interplay of strong spin-orbital coupling (SOC) and electronic correlations, such as Na_2IrO_3 etc. Recently, several authors have studied the phase diagram of the Kane-Mele-Hubbard (KMH) model[1, 2], which is used to describe the SOC and topological characters in these correlated systems. Generally, the semimetal is replaced by the Z_2 topological band insulator at finite SOC. At large electron correlation, long-range magnetic order breaks time-reversal invariance. The antiferromagnetic (AFM) Mott insulator phase exists and magnetic order is restricted to the xy plane in the presence of SOC.

The application of a magnetic field is a powerful approach to reveal hidden magnetic properties and their relationship with other order parameters. Some novel magnetic field may also exist with an external magnetic field. An external magnetic field breaks time reversal symmetry and it may lead to a Zeeman split to the spin motion. The interplay between the SOC and the Zeeman coupling may break the topological character. and it can also induce a nonlinear magnetic phase with electron Coulomb correlation. In the transition metal oxides such as Iridates system, the interplay between SOC and magnetic field with strong Coulomb correlation is an interesting topic.

In the present study, we investigate the KMH model under the Zeeman magnetic field. The phase diagrams with transverse and longitudinal magnetic fields are studied and the effects of competition between the magnetic field and SOC on local moments are discussed. This paper is organized as follows: we firstly describe the model Hamiltonian and theoretical approach, then the effects of external magnetic fields on the phase diagrams are discussed. The final section is devoted to the conclusion remarks.

2. Model Hamiltonian and Methods

The Kane-Mele model is proposed by Kane and Mele to describe the quantum spin Hall effect in graphene [1, 2]. To investigate the competition between the SOC and magnetic field, we employ the KMH model with external magnetic field, which can be written as follows:

$$H = \sum_{\langle ij\rangle\sigma}(t_{ij}c^{\dagger}_{i\sigma}c_{j\sigma} + h.c.) + U\sum_{i} n_{i\uparrow}n_{i\downarrow} - \frac{1}{2}\sum_{i}\mathbf{h}\cdot\mathbf{S}_i + i\lambda_{SO}\sum_{\langle\langle ij\rangle\rangle} v_{ij}c^{\dagger}_{i}\sigma^{z}c_j \quad (1)$$

where $c^{\dagger}_{i\sigma}$ and $c_{i\sigma}$ are the creation and annihilation operators for an electron of spin σ at site i. $n_{i\sigma}$ is the corresponding occupation number operator. σ^z is a Pauli matrix. Here v_{ij} equals ± 1 for sublattices A and B. λ_{SO} is the second-neighbor SOC strength. U is the Coulomb repulsion and $\mathbf{h}$ is an external magnetic field.

To treat the Coulomb correlation with arbitrary external field and spin-orbital coupling, the rotationally invariant slave boson (RISB) method is adopted[3, 4]. We introduce six slave bosons e, p_σ, p_x, p_y and b for empty, singly and double occupations, respectively. The enlarged local Hilbert space in the RISB representation is expressed by using these slave bosons and quasi-particle fermions f_σ. To focus the solution in the physical subspace, four constrains (including normalization, particle and spin constraints) given by Wölfle et al.[3] are introduced. Projecting the KMH model into the RISB representation, the Hamiltonian thus becomes

$$\underline{H} = \frac{1}{N}\sum_{aa'k\sigma\sigma'\sigma_{\prime\prime}}(R_{a\sigma\sigma'}\gamma_{a\sigma,a'\sigma''}(k)R_{a'\sigma'\sigma''} + (\varepsilon_0 - \mu)\delta_{\sigma\sigma''})\langle f^{\dagger}_{ak\sigma}f_{a'k\sigma''}\rangle + Ub^2 \quad (2)$$

where ε_0 and μ are the energy level and chemical potential. $R_{a\sigma\sigma'}$ and $R_{a'\sigma'\sigma''}$ are the elements of the renormalization matrix[3]. The dispersion $\gamma_{a\sigma a'\sigma''}(k)$ is written as

$$\gamma_{a\sigma,a'\sigma''}(k) = \begin{cases} -4\sigma v_a\lambda_{SO}\gamma_{SO}(k) + (\varepsilon_0 - \mu) - \frac{1}{2}\sigma h_z & \text{if } a = a' \text{ and } \sigma = \sigma'' \\ t\gamma_0(k)(1-\delta_{aa'}) + \frac{1}{2}(h_x + ih_y)\delta_{aa'} & \text{if } a \neq a' \text{ or } \sigma \neq \sigma'' \\ t\gamma_0^{\dagger}(k)(1-\delta_{aa'}) + \frac{1}{2}(h_x - ih_y)\delta_{aa'} & \text{if } a \neq a' \text{ or } \sigma \neq \sigma'' \end{cases} \quad (3)$$

where

$$\gamma_{SO}(k) = 4cos(\frac{\sqrt{3}}{2}k_y)sin(\frac{1}{2}k_x) + 2sin(k_x) \quad (4)$$

$$\gamma_0(k) = 1 + 2cos(\frac{1}{2}k_x)cos(\frac{\sqrt{3}}{2}k_y) - 2icos(\frac{1}{2}k_x)sin(\frac{\sqrt{3}}{2}k_y) \quad (5)$$

We solve the self-consistency equations that minimize the Eq. 2 by using optimization method[5].

3. Numerical Results

Generically, the full $SU(2)$ symmetry is broken by SOC terms. Thus the interaction between the SOC and electron Coulomb correlation exhibits rich phases [6, 7, 8, 9, 10, 11]. The existing results show that the topological band insulator (TBI) phase is stabilized by finite SOC. At the critical electron correlation, a transition from TBI phase to in plane Néel AFM phase occurs[8]. At intermediate electron correlation, a quantum spin liquid phase is also proposed[12]. However, some subsequent results show the direct phase transition from simple metallic phase to the AFM insulator phase[13, 14]. Thus we neglect the spin liquid phase in our work since it can be destroyed by external magnetic field.

We have studied the phase diagram and get the similar results as that given by M. Hohenadler et al.[8] except that the spin liquid phase is excluded. Here we first plot the magnetic phase

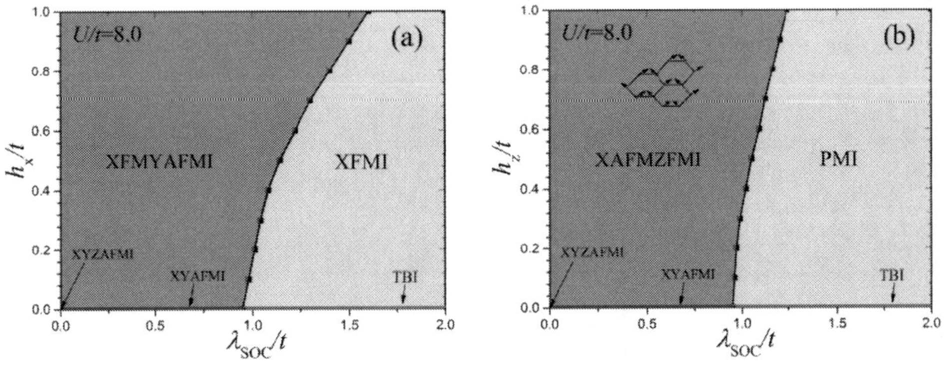

Figure 1. Ground-state phase diagram of the KMH model with transverse(a) and longitude(b) magnetic fields at $U/t = 8$. The phases are rotationally invariant Néel AFM insulator (XYZAFMI), in-plane antiferromagnetic insulator (XYAFMI), Z_2 topological band insulator (TBI), paramagnetic insulator (PMI), ferromagnetic insulator (XFMI) and canted AFM insulator (XAFMZFMI, XFMYAFMI) .

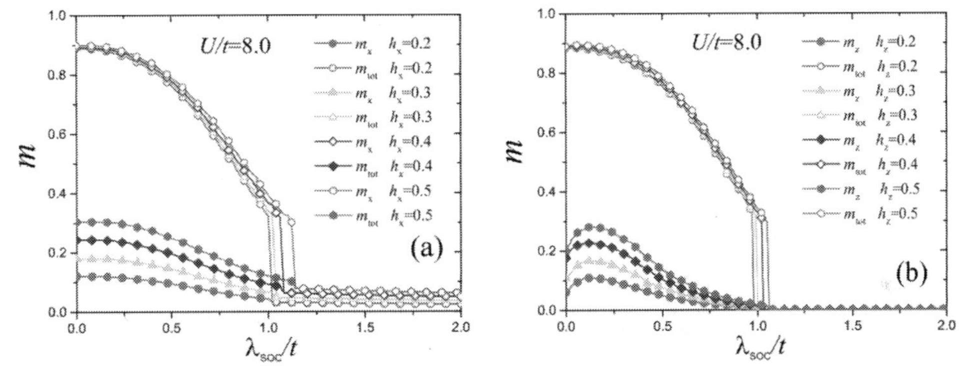

Figure 2. SOC dependences of the magnetic moments for different transverse(a) and longitude (b) magnetic fields at $U/t = 8$. m_{tot} is the total magnetic moment. m_x (m_z) is the corresponding $x(z)$ component of the local magnetic moments.

diagrams with transverse and longitude external magnetic fields in Fig.1. As expected, the ground states are Néel AFM phases at $U/t = 8$ in the absence of SOC and external magnetic field. A finite SOC induces an in-plane AFM phase due to the z direction AFM exchange coupling between the next nearest neighbors[8]. With further increasing SOC, a XYAFMI-TBI transition occurs, which has been proposed by many works[8, 15, 16]. With $h_x > 0$, the system translate into XFMYAFMI phase as shown in Fig.1(a). In this phase, the spin arranges AFM in y direction and ferromagnetic in x direction. With increasing SOC, the AFM component are suppressed completely and the system translates into ferromagnetic insulator phase with the spin arranges in x direction (XFMI). The phase diagram with finite h_z is also plotted in Fig.1(b). In the canted insulator phase XAFMZFMI, the spin arranges AFM in x direction and ferromagnetic in z direction. We find that the paramagnetic insulator phase appears with increasing SOC due to the competition between the external magnetic field and SOC.

The corresponding magnetic moments at $U/t = 8$ are plotted in Fig. 2. The total magnetic moments are nearly independent on the external field since the local spin has polarized absolutely due to the strong electron Coulomb correlation. But the component of the local magnetic moment along the external magnetic field increases considerably with increasing magnetic field. Due to the exchange interaction between next-nearest neighbours [8], the total magnetic moments and components along the external magnetic fields decrease with increasing SOC. As expected, the x component of the local magnetic magnetic moment is finite with the λ_{SO} large than critical value as shown in Fig. 2(a). From Fig. 2(b), we find that there is a peak of the z components of the local moments with increasing λ_{SO}, which only exists with longitudinal external magnetic field and strong electron correlation. The reason is that the in-plane AFM component is depressed by the finite λ_{SO}.

4. Summary

In summary, we have studied the effects of external magnetic field on the magnetic phase diagrams of KMH model. Our results show that the system translates into canted AFM state under external magnetic field, which is destroyed by large SOC. The effects of longitudinal and transverse external magnetic fields on the ground states are very different. Although the magnetism can be suppressed by large SOC under longitudinal magnetic field, a finite SOC favors the longitudinal component of the local magnetic moment.

5. Acknowledgments

This work was supported by the National Science Foundation of China under Grant no. 11204311, 11574315 and 11104148. Numerical calculations were performed at the Center for Computational Science of CASHIPS.

6. References

[1] C. L. Kane and E. J. Mele, *Phys. Rev. Lett.* **95** 146802 (2005).
[2] C. L. Kane and E. J. Mele, *Phys. Rev. Lett.* **95** 226801 (2005).
[3] T. Li, P. Wölfle and P. J. Hirschfeld, *Phys. Rev. B* **40** 6817 (1989).
[4] R. Frésard and P. Wölfle, *Int. J. Mod. Phys. B* **6** 685 (1992).
[5] Y. M. Quan, Q. W. Wang, D. Y. Liu, X. L. Yu, L. J. Zou, *Computer Physics Communications* **191**, 90-99 (2015).
[6] S. Rachel and K. Le Hur, *Phys. Rev. B* **82** 075106 (2010).
[7] M. Hohenadler, T. C. Lang and F. F. Assaad, *Phys. Rev. Lett.* **106** 100403 (2011).
[8] M. Hohenadler, Z. Y. Meng, T. C. Lang, S. Wessel, A. Muramatsu, and F. F. Assaad, *Phys. Rev. B* **85** 115132 (2012).
[9] D. Zheng, G. M. Zhang and C. Wu, *Phys. Rev. B* **84** 205121 (2011).
[10] S. L. Yu, X. C. Xie and J. X. Li, *Phys. Rev. Lett.* **107** 010401 (2011).
[11] W. Wu, S. Rachel, W.-M. Liu and K. Le Hur, *Phys. Rev. B* **85** 205102 (2012).
[12] Z. Y. Meng, T. C. Lang, S. Wessel, F. F. Assaad, and A. Muramatsu, *Nature* **464** 847 (2010).
[13] S. Sorella, Y. Otsuka, and S. Yunoki, *Sci. Rep.* **2** 992 (2012).
[14] M. Ebato, T. Kaneko, and Y. Ohta, *J.Phys.Soc.* **84** 044714 (2015).
[15] S. Yu, X. C. Xie, and J. X. Li, *Phys. Rev. Lett.* **107** 010401 (2011).
[16] J. Reuther, R. Thomale, and S. Rachel, *Phys. Rev. B* **86** 155127 (2012).

Possible three-dimensional superconducting properties of strongly correlated honeycomb compound $In_3Cu_2VO_9$

Shi-qing Jia[1], Qing-wei Wang[2], Jing liu[2], Liang-Jian Zou[2]

[1] School of Physics and Materials Science, Anhui University, Hefei 230039, China, and Key Laboratory of Materials Physics, Institute of Solid State Physics, Chinese Academy of Sciences, Hefei 230031, P.R. China

[2] Key Laboratory of Materials Physics, Institute of Solid State Physics, Chinese Academy of Sciences, Hefei 230031, P.R. China

E-mail: zou@theory.issp.ac.cn

Abstract. We present the zero- and finite-temperature superconducting properties of a modified $t - J - J_c$ model on three-dimensional honeycomb lattice by employing the strongly correlated *Slave Boson* mean-field theory, focusing on the low energy physics in $In_3Cu_2VO_9$. We find that the highest superconducting transition temperature $T_c \approx 4$ K occurs at doping hole concentration $\delta = 0.03$. The superconducting phase disappears at doping concentration of $\delta = 0.06$. The superconducting ground state in doped $In_3Cu_2VO_9$ supports a $d_{x^2-y^2} + id_{xy}$ pairing symmetry when the interlayer coupling J_c increases from 1.2 to 5.2 meV. Our theoretical results suggest that doped $In_3Cu_2VO_9$ is a possible superconducting honeycomb system.

1. Introduction

Unconventional high-temperature superconductivity (SC) in two-dimensional honeycomb lattice has attracted great attention in recent years, since its low-energy quasi-particles in normal state are Dirac particles with relativistic linear spectrum. However, such candidate compounds are rather rare, especially in strongly correlated systems with strong spin fluctuations. Layered honeycomb copper oxide $In_3Cu_2VO_9$ synthesized by Möller *et al.* behaves very similar electronic states to those in high-T_c cuprates. It lies in strongly correlated limit with ($U/t \gg 1$), the Cu-3d electron is half filling with $3d^9$ and $S = 1/2$, and Cu-O-Cu forms a superexchange antiferromagnetic (AFM) coupling [1]. A recent work by Liu *et al.* presented the tight-bonding parameters t_{ij} and the spin coupling strengths J_{ij} of undoped $In_3Cu_2VO_9$ [2], demonstrating that it is a strongly correlated system. We also showed that the ground state of such a layered honeycomb compound is spin disordered [3], implying a considerable spin fluctuations. This arises a great interesting problem: whether could doped $In_3Cu_2VO_9$ exhibit high-T_c SC like doped cuprates ?

Until now, there has not been any experimental evidence for SC phase upon doping in $In_3Cu_2VO_9$, since the electron or hole carriers are difficult to be doped into the compound, no insulator-metal transition occurs [4], not to mention the metal-SC transition. Recently, the development of liquid or solid gating technology makes it possible to tune the carrier concentration in strongly correlated compounds, appealing for the exploration of the SC

IOP Conf. Series: Journal of Physics: Conf. Series **827** (2017) 012010 doi:10.1088/1742-6596/827/1/012010

properties of $In_3Cu_2VO_9$. Meanwhile, the SC properties in monolayer honeycomb lattice, such as in graphene, were extensively studied by many theorists, for example, see Ref.[5, 6]. However, strongly correlated $In_3Cu_2VO_9$ is considerably different from weakly interacting graphene and strongly correlated cuprates. On the one hand, no AFM long-range order (LRO) and AFM transition were observed in the specific heat of undoped $In_3Cu_2VO_9$ down to 2 K [1]. Instead, its magnetic susceptibility and specific heat exibihit a very broad bump over a wide temperature range from 50 K to 300 K [4], implying a possibility of the spin disorder or liquid insulating ground state, rather than an AFM insulating ground state. On the other hand, according to Ref.[2], there are two kinds of magnetic frustrations in $In_3Cu_2VO_9$: one is from the intralayer next-nearest-neighbour (NNN) AFM spin coupling, the other comes from the interlayer nearest-neighbour (NN) spin coupling. The interlayer magnetic frustration might stabilize the exotic spin disorder phase and contribute to the SC phase in finite doping.

Thus, it is essential to explore the role of interlayer frustration on the SC ground state and the SC symmetry of the effective t-J_1-J_2-J_c model in a three-dimensional (3D) system, so as to mimic the low-energy physics of doped $In_3Cu_2VO_9$. In addition, very few work has come to the doped hole concentration δ dependence of the transition temperature T_c in frustrated honeycomb lattice. In the rest of this paper, we first outline the model Hamiltonian and the *Slave boson* mean-field theory (*SBMFT*) in Sec.2, and then present the parameter and doping dependences and pairing symmetry of the SC properties in Sec.3, and the final section is devoted to the summary.

2. Model Hamiltonian and Slave Boson Method

In accordance with Ref.[2], the low-energy physics of doped $In_3Cu_2VO_9$ is dominated by the kinetic energy and the AFM spin interactions in the honeycomb planes of copper ions, *i.e.* a generalized $t-J$ model. The model Hamiltonian thus reads as:

$$H = -\sum_{ij,\sigma} t_{ij}\left(\hat{f}^\dagger_{i\sigma}\hat{f}_{j\sigma} + h.c.\right) + \sum_{ij} J_{ij}\left(\mathbf{S}_i\cdot\mathbf{S}_j - \frac{1}{4}n_i n_j\right) \quad (1)$$

where nonzero t_{ij} are the intralayer NN, NNN and interlayer NN hopping integrals t_1, t_2, and t_c; and finite J_{ij} are the intralayer NN, NNN and interlayer NN AFM spin interactions J_1, J_2 and J_c, respectively. Due to the large-U limit, the high-energy double occupation state should be projected out in the electron projection operator $\hat{f}_{i\sigma}$, so the Coleman-Barnes' SBMFT [7] is a suitable technique to treat such a strongly correlated Hamiltonian.

The essence of the *SBMFT* is to project the electron operator $\hat{f}_{i\sigma}$ as a product of holon boson $h_{i\sigma}$ of the charge and spinon fermion g_i of the spin,

$$\hat{f}^\dagger_{i\sigma} = g_i h^\dagger_{i\sigma}, \quad \hat{f}_{i\sigma} = h_{i\sigma} g^\dagger_i. \quad (2)$$

with the constraint condition

$$g^\dagger_i g_i + \sum_\sigma h^\dagger_{i\sigma} h_{i\sigma} = 1. \quad (3)$$

on each sublattice site. The holons and spinons satisfy commutation and anticommutation relation, respectively. Within the *slave boson* framework, the projected generalized $t-J$ model Hamiltonian is rewritten as,

$$\begin{aligned} H = & -\sum_{ij,\sigma} t_{ij}\left(g_i g^\dagger_j h^\dagger_{i\sigma} h_{j\sigma} + h.c.\right) - \mu\sum_{i\sigma} h^\dagger_{i\sigma} h_{i\sigma} + \sum_i \lambda_i\left(g^\dagger_i g_i + \sum_\sigma h^\dagger_{i\sigma} h_{i\sigma} - 1\right) \\ & + \frac{(1+\delta)^2}{2}\sum_{ij} J_{ij}\left(h^\dagger_{i\sigma} h^\dagger_{j\bar{\sigma}} h_{j\bar{\sigma}} h_{i\sigma} + h^\dagger_{i\sigma} h^\dagger_{j\bar{\sigma}} h_{i\bar{\sigma}} h_{j\sigma}\right) \end{aligned} \quad (4)$$

IOP Conf. Series: Journal of Physics: Conf. Series **827** (2017) 012010 doi:10.1088/1742-6596/827/1/012010

where μ is the chemical potential and λ is the *Lagrangian* multiplier. In according with the results of the *first-principles* electronic structures calculations, we take t_1, t_2, t_c, J_1, and J_2 are -180, -24.1, -48.9, 16.2, and 0.3 meV, respectively [2]. Note that the interlayer NN coupling J_c was estimated from the AFM ordered phases, we expect that it was well underestimated. In this paper we take $J_c = 4.2\ meV$ [3].

To deal with such a 3D strongly correlated and frustrated model Hamiltonian Eq.(4), we first adopt the *SBMFT* to explore the pairing symmetry of SC phases. Within the *SBMFT* technique, we define the order parameters representing the resonating valence bond (RVB) spinon singlet $\Delta_{\alpha\beta r}$ and the coherent motion of spinons $P_{\alpha\beta}$ and motion of holons $Q_{\alpha\beta}$, respectively. Here α, β represent the four sublattices in 3D $In_3Cu_2VO_9$.

$$\begin{aligned}
\Delta_{\alpha\beta r} &= -\frac{J_{1(c)}}{2}\langle h_{\beta i+r\downarrow}h_{\alpha i\uparrow} - h_{\beta i+r\uparrow}h_{\alpha i\downarrow}\rangle, \\
\Delta_{\alpha\alpha r} &= -\frac{J_2}{2}\langle h_{\alpha i+r\downarrow}h_{\alpha i\uparrow} - h_{\alpha i+r\uparrow}h_{\alpha i\downarrow}\rangle, \\
P_{\alpha\beta} &= \sum_\sigma \langle h^\dagger_{\alpha i\sigma}h_{\beta i+r\sigma}\rangle, \quad Q_{\alpha\beta} = \langle g^\dagger_{\alpha i}\, g_{\beta i+r}\rangle.
\end{aligned} \tag{5}$$

here r is the NN or NNN direction vector, the corresponding coordination numbers z_1, z_2, and z_c are 3, 6, and 2 for intralayer NN, intralayer NNN and interlayer NN, respectively. The *Fourier* transformation of these RVB gap functions are then: $\Delta_{\alpha\beta}(\mathrm{k})=\sum_r \Delta_{\alpha\beta r}\ e^{ik\cdot r}$, and $\Delta_{\alpha\alpha}(\mathrm{k})=\sum_r \Delta_{\alpha\alpha r}\cos(k\cdot r)$.

At $T = 0$ K, the holon hopping order parameters $Q_{\alpha\beta} \approx \delta$. Thus we could rewrite the Hamiltonian Eq.(4) in the *SBMFT* approach in the momentum space.

$$\begin{aligned}
H^{MF} = &- \sum_{\alpha\beta m,k\sigma} z_m\gamma_{\alpha\beta}\left(t_m\delta + \frac{1}{4}J_m P^*_{\alpha\beta}\right) h^\dagger_{\alpha k\sigma}h_{\beta k\sigma} \\
&+ \sum_{\alpha\beta,k}\left(\Delta_{\alpha\beta}(k)h^\dagger_{\alpha k\uparrow}h^\dagger_{\beta -k\downarrow} + \Delta^*_{\alpha\beta}(k)h^\dagger_{\beta k\uparrow}h^\dagger_{\alpha -k\downarrow}\right) \\
&+ \sum_{\alpha k\sigma}\left[\lambda - \mu - z_2\gamma_{\alpha\alpha}\left(t_2\delta + \frac{1}{4}J_2 P_{\alpha\alpha}\right) - \frac{n}{4}\Sigma_\nu J_\nu z_\nu\right] h^\dagger_{\alpha k\sigma}h_{\alpha k\sigma} \\
&+ \sum_{\alpha k}\Delta_{\alpha\alpha}(k)h^\dagger_{\alpha k\uparrow}h^\dagger_{\alpha -k\downarrow} + H.c. + E_0
\end{aligned} \tag{6}$$

with the constant energy E_0 defined as

$$E_0 = 2N\sum_{\alpha\beta r\nu}\frac{|\Delta_{\alpha\beta r}|^2}{J'_\nu} + N\sum_{\alpha\beta\nu}\frac{z_\nu J'_\nu}{4}\left(n^2 + |P_{\alpha\beta}|^2\right) - 4N(\lambda-\mu)n \tag{7}$$

where the subscript m runs over 1 and c, and ν over 1, 2, and c; $J'_\nu = J_\nu(1+\delta)^2$. $\gamma_{\alpha\beta}$ are the geometrical structure factors of the intralayer NN, interlayer NN and intralayer NNN spins.

Diagonalize the *SBMFT* Hamiltonian H^{MF} by employing a complex 8×8 Bogoliubov transformation [7], we obtain eight energy dispersion relations $\omega_{ku\sigma}$ of the fermion excitations. Thus we achieve the free energy $F = -k_B T \ln Tr\left(e^{-\beta H^{MF}}\right)$ as follows:

$$F = -k_B T\sum_{ku\sigma}\ln\left[1 + \exp\left(-\beta\omega_{ku\sigma}\right)\right] + E_0 \tag{8}$$

Then minimizing the free energy with respect to the RVB order parameters $\Delta_{\alpha\beta r}$, spinon motion order parameters $P_{\alpha\beta}$, the chemical potential μ and the *Lagrangian* multiplier λ, one arrives the

IOP Conf. Series: Journal of Physics: Conf. Series **827** (2017) 012010 doi:10.1088/1742-6596/827/1/012010

saddle points of free energy minimum. The saddle point satisfies the following self-consistent equations,

$$\begin{aligned}
\delta &= \frac{1}{4N}\sum_{ku\sigma}\frac{\partial \omega_{ku\sigma}}{\partial \mu}\frac{1}{1+\exp\left(\beta\omega_{ku\sigma}\right)} \\
\Delta_{\alpha\beta r} &= -\frac{J_{1(c)}}{2N}\sum_{ku\sigma}\frac{\partial \omega_{ku\sigma}}{\partial \Delta^{*}_{\alpha\beta r}}\frac{1}{1+\exp\left(\beta\omega_{ku\sigma}\right)} \\
\Delta_{\alpha\alpha r} &= -\frac{J_{2}}{2N}\sum_{ku\sigma}\frac{\partial \omega_{ku\sigma}}{\partial \Delta^{*}_{\alpha\alpha r}}\frac{1}{1+\exp\left(\beta\omega_{ku\sigma}\right)} \\
P_{\alpha\beta} &= -\frac{4}{3(2)J_{1(c)}N}\sum_{ku\sigma}\frac{\partial \omega_{ku\sigma}}{\partial P^{*}_{\alpha\beta}}\frac{1}{1+\exp\left(\beta\omega_{ku\sigma}\right)} \\
P_{\alpha\alpha} &= -\frac{4}{3J_{2}N}\sum_{ku\sigma}\frac{\partial \omega_{ku\sigma}}{\partial P^{*}_{\alpha\alpha}}\frac{1}{1+\exp\left(\beta\omega_{ku\sigma}\right)} \qquad (9)
\end{aligned}$$

Solving these self-consistent equations, we could obtain the doping hole concentration δ and interlayer NN J_c dependences of moduli of the order parameters and the SC phases' symmetry in the ground state. We set the moduli of the intralayer NN, intralayer NNN and interlayer NN RVB order parameters as $|\Delta_{\alpha\beta r}| = \Delta_1$, $|\Delta_{\alpha\alpha r}| = \Delta_2$, and $|\Delta_{\alpha\beta r}|=\Delta_3$ or Δ_4, respectively.

3. Numerical Results and Phase Diagrams

Performing the numerical calculation to Eq.(9), we could explore the SC groundstate properties of the 3D $In_3Cu_2VO_9$ in the framework of the *SBMFT*.

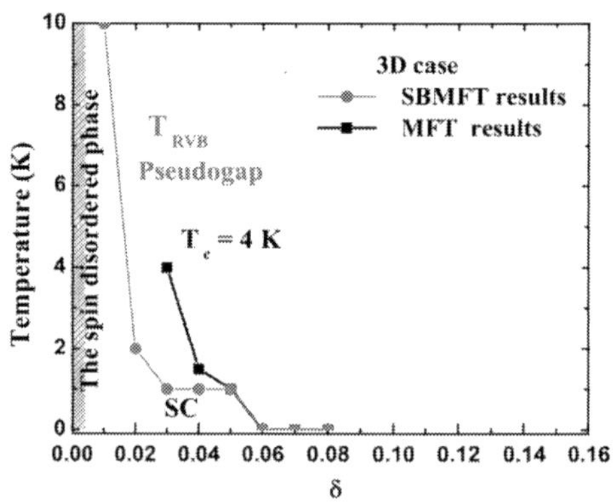

Figure 1. (Color online) Dependence of the transition temperature T_{RVB} on hole doping concentration δ obtained by *SBMFT* and mean-field solution T_c in 3D $In_3Cu_2VO_9$.

Fig.1 displays the hole doping concentration dependence of the SC transition temperature obtained by the *SBMFT*, the conventional mean field solution is also plotted for comparison. It shows that the mean-field T_c decreases with the increase of hole doping concentration. The highest SC transition temperature $T_c \approx 4$ K occurs at optimized doping concentration $\delta = 0.03$, and the SC phase disappears at $\delta = 0.06$ ($\ll 1$) in the mean-field approximation, as shown in Fig.1. In the *SBMFT* solutions, we obtain the RVB state transition temperature T_{RVB} which indicates the formation of the singlet spin pairing, as shown the red line in Fig.1. It also decreases

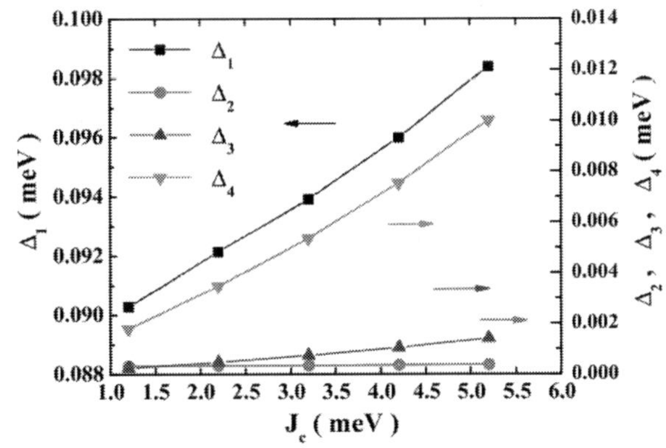

Figure 2. (Color online) Dependence of the RVB order parameters $\Delta_{1\sim4}$ on interlayer NN spin coupling J_c at optimized doping $\delta = 0.03$ in 3D $In_3Cu_2VO_9$.

with the increase of δ. Then T_{RVB} can also explain the SC pseudogap and the SC phase T_c should have been below them.

We plot the J_c dependence of the RVB order parameters Δ at optimized doping $\delta = 0.03$ in Fig.2. From Fig.2, we find that the moduli of the intralayer NN RVB order parameters Δ_1 are much larger than the intralayer NNN and interlayer NN ones, Δ_2 and Δ_3 or Δ_4 because $J_1 \gg J_2, J_c$. Therefore, the SC pairing symmetry mainly depends on the intralayer NN gap functions $\Delta_{\alpha\beta}(\mathrm{k})$. With the increase of J_c, all the RVB order parameters $\Delta_{1\sim4}$ increase monotonically, as shown in Fig.2, suggesting that the stronger the interlayer spin fluctuations are, the larger the SC pairing forces are.

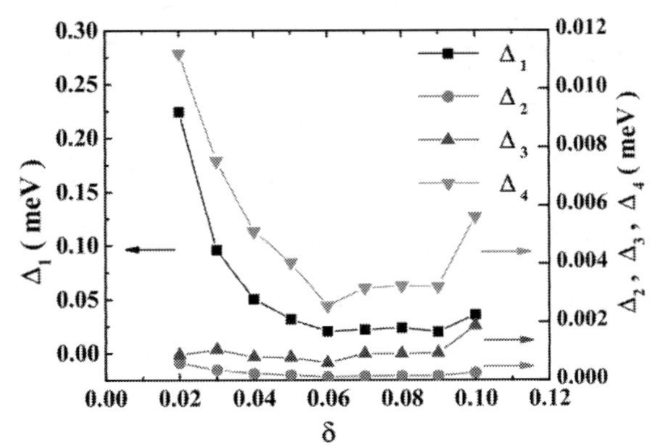

Figure 3. (Color online) Hole doping concentration dependence of the moduli of the RVB order parameters Δ for $J_c = 4.2\ meV$ in 3D $In_3Cu_2VO_9$.

To get insight into the evolution of the SC pairing on the doping concentration, we plot the dependence of the RVB order parameters on the doping concentration of the holes in Fig.3. From Fig.3, we find that with the increase of doping hole δ, all the moduli of the RVB order parameters $\Delta_1, \Delta_2, \Delta_3, \Delta_4$ decrease with the increase of δ, until $\delta = 0.06$. After $\delta > 0.06$, the situation is rather abnormal. Since these order parameters $\Delta_{1\sim4}$ slightly increase with the

IOP Conf. Series: Journal of Physics: Conf. Series **827** (2017) 012010 doi:10.1088/1742-6596/827/1/012010

increase of doping concentration. This seems to show that there exists a second SC phase, which deserves further study in the near future.

The SC paring symmetry is an important property for the present unconventional SC in doped $In_3Cu_2VO_9$. Our analysis shows that the intralayer NN SC gap functions $\Delta_{\alpha\beta}(\mathrm{k})$ support the $d_{x^2-y^2}+id_{xy}$ symmetry, since the mean-field solution $\Delta_{\alpha\beta}(k) = \Delta_1 e^{i2\pi(n_r-1)/3}$, here n_r is the serial number of r. This confirms the C_3 symmetry of the SC pairing order parameters [6]. The intralayer NNN SC order parameters $\Delta_{\alpha\alpha}(k) = \Delta_2 e^{i2\pi(n_r-1)/3}$, which is also consistent with the C_3 symmetry. Thus the intralayer NNN gap function $\Delta_{\alpha\alpha}(\mathrm{k})$ also satisfies the $d_{x^2-y^2}+id_{xy}$ symmetry. Meanwhile, the interlayer NN SC order parameters are real. Therefore, totally speaking, doped $In_3Cu_2VO_9$ supports a SC ground state with a $d_{x^2-y^2}+id_{xy}$ pairing symmetry.

4. Conclusion

In summary, we have shown that in the presence of interlayer spin frustrations and quantum fluctuations in doped 3D $In_3Cu_2VO_9$, the possible highest SC transition temperature $T_c \approx 4$ K at optimum doping hole concentration $\delta = 0.03$ and the SC phase disappears at $\delta = 0.06$. The interlayer NN exchange coupling is beneficial to the formation of SC pairing, since the SC order parameters obtained by the mean field method increase with lifting J_c, though its enhancement to T_c is not significant. In contrast, our *Slave boson* solution shows that the intralayer NN RVB order parameters Δ_1 increase almost linearly with the lifting of J_c, and the holon hopping parameters $P_{\alpha\beta}$ decrease almost linearly. Therefore, the interlayer spin frustrations and quantum fluctuations favor to the possible superconductivity in doped $In_3Cu_2VO_9$. We expect more experiments to verify our prediction.

Acknowledgments

This work was supported by the National Science Foundation of China under Grant no. 11474287 and 11274310. Numerical calculations were performed at the Center for Computational Science of CASHIPS.

References

[1] A Möller, U Löw, T Taetz, M Kriener, G André, F Damay, O Heyer, M Braden and J A Mydosh 2008 *Phys. Rev. B* **78** 024420
[2] D-Y Liu, Y Guo, X-L Zhang, J-L Wang, Z Zeng, H-Q Lin and L-J Zou 2013 *Europhys. Lett.* **103** 47010
[3] S-Q Jia, Q-W Wang, X-L Yu and L-J Zou 2017 *AIP Adv.* **7** 055825
[4] Y-J Yan, Z-Y Li, T Zhang, X-G Luo, G-J Ye, Z-J Xiang, P Cheng, L-J Zou and X-H Chen 2012 *Phys. Rev. B* **85** 085102
[5] S Pathak, V B Shenoy and G Baskaran 2010 *Phys. Rev. B* **81** 085431
[6] A M Black-Schaffer and S Doniach 2007 *Phys. Rev. B* **75** 134512
[7] Z-Z Li 2002 *Solid State Theory* (Beijing: Higher Education Press) chapter 11 pp 450-468

IOP Conf. Series: Journal of Physics: Conf. Series **827** (2017) 012011 doi:10.1088/1742-6596/827/1/012011

Spin and valley transport in the ferromagnetic MoS_2 junctions subjected by the gate voltage

P. Ye[1], R.Y. Yuan[1], Y. Y. Xia[1], X. Zhao[2]

[1]Center for Theoretical Physics, Deparment of Physics, Capital Normal University, Bejing 100048, China.

[2]Department of Physics and State Key Laboratory of Low-Dimensional Quantum Physics, Tsinghua University, Bejing 100084, China

Electronic address: yuanry@cnu.edu.cn

Abstract. In this paper, we have studied valley and spin-polarized transport with different gate voltage in ferromagnetic MoS_2 junctions. The results show that, the valley and spin transport through the junctions has a large oscillation. In particular, a fully spin polarized current can be put out by effectively tuning the gate voltage U. Moreover, with increasing of U, the scope of the valley and spin polarization can be greatly expanded. These findings indicate the structure is a considerable candidate for the spintronics or valleytronics device.

1. Introduction

The monolayer (ML) MoS_2, a typical two-dimensional (2D) material, have attracted growing attention. The ML MoS_2 has a relatively large direct energy gap compared to graphene[2], possessing high carrier mobility, large thermal stability and good compatibility with standard semiconductor manufacturing[3]. These remarkable physical properties make ML MoS_2 as potential applications in next-generation nanoelectronic devices. Moreover, ML MoS_2 has strong spin-orbit coupling (SOC) absent in graphene[4,11-12], which originated from the d orbitals of the heavy metal atoms. In addition, akin to graphene, the conduction and valence-band edges in monolayer MoS_2 are located at the corners (K points) of the 2D hexagonal Brillouin zone, and the two inequivalent valleys constitute a binary index for low energy carriers. Besides, the inversion symmetry in monolayer MoS_2 breaking leads to the valley Hall effect where carriers from different valleys flow to opposite transverse edges when an in-plane electric field is applied[1,5-6]. It is extremely desirable for the spin and valley-based electronic and valley control applications of monolayer MoS_2[7,9,13].

In this letter, we consider a system where the ferromagnetic ML MoS_2 is placed in the center, both sides being nonferromagnetic. A voltage gate is placed on the top of the ferromagnetic ML MoS_2. The exchange field can be induced by the magnetic proximity effect by depositing an array of ferromagnetic dysprosium strips.

2. Theory and Formula

Now let us explain the formulation. The fermions around the Fermi level in MoS_2 obey a massless relativistic Dirac equation. The Hamiltonian is given by[8]

$$H = \hbar v_F(\tau_z k_x \sigma_x + k_y \sigma_y) + \frac{\Delta}{2}\sigma_z - \lambda\tau_z \frac{\sigma_z - 1}{2} s_z + U + s_z h, \quad (1)$$

U by a voltage gate caused is regarded as in the ferromagnetic region and 0 otherwise, and L is the ferromagnetic region width. h is the exchange field added in the ferromagnetic region. Where $\tau_z = \pm 1$ stands for K and K' valleys, respectively. σ_{xyz} represents the Pauli matrix in sublattice space. $s_z = \pm 1$ denotes the spin-up ↑ and spin-down ↓ components. The Fermi velocity $v = 5.3 \times 10^5$ m·s^{-1}, λ =37.5 meV is the spin splitting induced by the SOC, and Δ = 833 meV is the band gap. h is the exchange field added in the ferromagnetic region. According to Eq. (1), k_x' in the ferromagnetic region is given by

$$k_x' = \{[E - U - (\tau_z s_z \lambda + s_z h)]^2 - (\Delta - \tau_z s_z \lambda)^2\}^{\frac{1}{2}} \cdot (\hbar v_F)^{-1}. \quad (2)$$

In the nonferromagnetic regions, the wave vector k_x can be similarly obtained by substituting h =0, U= 0 according to Eq. (2). By solving the equation HΨ(x, y) = EΨ(x, y), and then the spin- and valley-resolved transmission probability can be analytically obtained

$$T_{\tau_z s_z} = \left| \frac{4k_x k_x' E_M E_F \exp(-ik_x L)}{\chi\xi} \right|^2$$

$$\xi = [-E_M^2 k'^2 + 2E_M E_F(k_x k_x' + k_y{}^2) - E_F^2 k^2]\exp(ik_x' L)$$

$$\chi = [E_M^2 k'^2 + 2E_M E_F(k_x k_x' - k_y^2) + E_F^2 k^2]\exp(-ik_x' L)$$

$$E_M = E - \Delta \quad E_F = E - U - \Delta + s_z h. \quad (3)$$

To evaluate the electron valley- and spin-resolved polarization, we define total tunneling transmission difference $T_{K(K')}$ and $T_{\uparrow(\downarrow)}$, which read

$$T_{K(K')} = \frac{T_{K(K')\uparrow} + T_{K(K')\downarrow}}{2} \qquad T_{\uparrow(\downarrow)} = \frac{T_{K\uparrow(\downarrow)} + T_{K'\uparrow(\downarrow)}}{2}.$$

By setting $k_x = k\cos(\phi)$ and $k_y = k\sin(\phi)$, we define the normalized valley and spin resolved conductance

$$G_{\tau s} = \frac{1}{2}\int_{-\frac{\pi}{2}}^{\frac{\pi}{2}} |t_{\tau s}|^2 \cos(\phi) d\phi, \quad (4)$$

then we obtain the valley and spin polarization P_v and P_s

$$P_v = \frac{(G_{K\uparrow} + G_{K\downarrow}) - (G_{K'\uparrow} + G_{K'\downarrow})}{G_{K\uparrow} + G_{K\downarrow} + G_{K'\uparrow} + G_{K'\downarrow}} \qquad P_s = \frac{(G_{K\uparrow} + G_{K'\uparrow}) - (G_{K'\downarrow} + G_{K\downarrow})}{G_{K\uparrow} + G_{K\downarrow} + G_{K'\uparrow} + G_{K'\downarrow}}$$

3. Result and discussion

First, to demonstrate the discrepancy of valley and spin transmission, we have calculated the corresponding transmission coefficients of $T_{K(K')}$ and $T_{\uparrow(\downarrow)}$. Figure 1 shows these transmission coefficients versus the incident energy E with different U, where figure 1(a) and figure 1(b) correspond to the $T_{K(K')}$ and $T_{\uparrow(\downarrow)}$, respectively. The structure parameters are chosen to be L=7nm, $\Delta = E_0$, $h = 0.1E_0$ and $k_y = 0$, From figure 1(a), when U=0.5E$_0$, we can obviously see that the transmission of the K' electrons and K electrons in lower energy region is blocked by the barrier. Importantly, the critical energy of K' electrons is larger than that of the K electrons. This interesting feature shows that the two valleys have different behavior, leading to valley-polarized transport which is shown in figure 2(a). However, the energy range is extremely narrow. Furthermore, the difference transmission of between K electrons

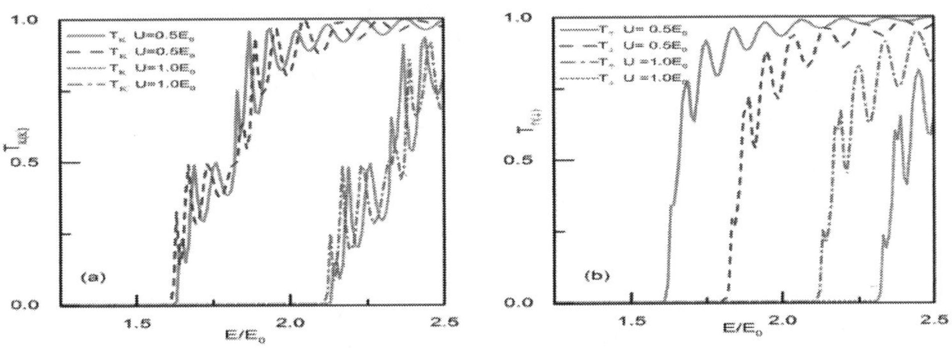

Figure 1. (a) Valley transmission and (b) spin transmission as a function of incident energy E under the manipulation of various U

and K' electrons in high energy region gradually disappears. For U=1.0E$_0$, oscillation of the transmission of K electrons and K' electrons are more prominent than U=0.5E$_0$. In additional, the increase of U can expand the range of valley-polarized transport (see the figure 2(a)), owing to the external barrier U making the conductance band edge shifting upwards. Next, we show the spin resolved transmission and spin-polarized conductance for different U. We observe that the difference between transmission of spin-up states and transmission of spin-down state are most apparent in figure1(b). In comparison to the case of the valley resolved transmission, we can see that spin-up state is stronger than that for spin-down one in a wider energy range. Especially, from 1.55E$_0$ to 1.65E$_0$, spin-down states are absolutely suppressed while spin-up states are filtered out, which results in a fully spin-polarized transport (see the figure 2(b)). Furthermore, it is noted that, for U=0.5E$_0$ in figure 2(b), a fully spin polarization conductance is obtained in 1.2E$_0$ to 1.65E$_0$ energy range. However, for U=1.0E$_0$, it can be seen that spin polarization can be achieved while absent of spin filtering in the same energy range. From 1.65E$_0$ to 2.2E$_0$, spin-polarized conductance value is up to 1, and this means the spin-polarized conductance can be effectively tuned by the gate voltage.

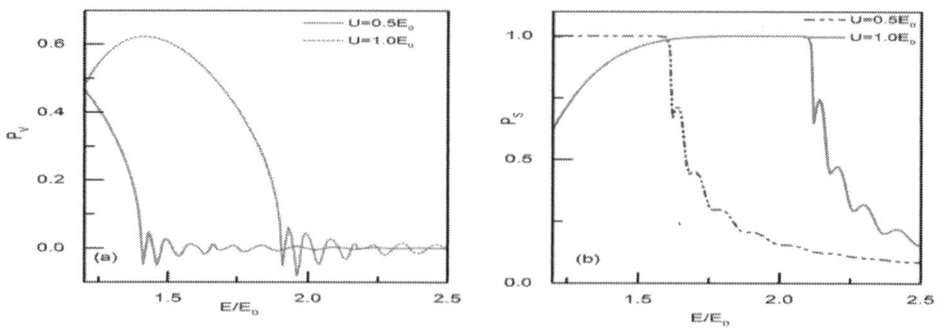

Figure 2. (a) Valley-polarized conductance and (b) spin-polarized conductance as a function of incident energy E under the manipulation of various U.

4. Conclusion

In summary, we have studied the valley and spin transport in ferromagnetic MoS_2 junctions. The results showed that a fully spin polarized conductance can be obtained by modulation of the gate voltage. At the same time, valley-polarized transmissions in a small energy area can also be achieved. But the value is only about up to 0.6 by tunneling the gate voltage. These results provide an avenue for the electrical control of valley-and spin-polarized transport in MoS_2-based devices.

Acknowledgments

This work is supported by NSFC under grantsNo.11574173 Open Research Fund Program of the State Key Laboratory of Low Dimensional Quantum Physics under grants No.KF201510.

References

[1] Xiao D, Liu G B, Feng W, Xu X and Yao W, *Phys.Rev. Lett.* 2012, **108**, 196802.
[2] Ellis J K, Lucero M J and Scuseria G E, *Appl. Phys. Lett.* 2011, **99**, 261908.
[3] Radisavljevic B, Radenovic A, Brivio, J, *Nat. Nanotechnol.* 2011, **6**, 147-150.
[4] Sarma S D, Adam S, Hwang E H and Rossi E, *Rev. Mod. Phys.* 2011, **83**, 407-470.
[5] Zhu Z Y, Cheng Y C and Schwingenschlogli U, *Phy. Rev. B* 2011, **84**, 153402.
[6] Cheiwchanchamnangij T, Lambrecht W R L, *Phys. Rev. B* 2012, **85**, 205302.
[7] Mak K F, He K, Shan J, Heinz T F, *Nat. Nanotechnol.*2012, **7**, 494−498.
[8] Lu H Z, Yao W, Xiao D, Shen S Q, *Phys. Rev.Lett.* 2013, **110**, 016806.
[9] Cao T, Wang G, Han W, Ye H, Zhu C, Shi, J, Niu Q, Tan P, Wang E, *Nat. Commum.* 2012, **3**, 887.
[10] Zeng H, Dai J, Yao W, Xiao D and Cui X, *Nat. Nanotechnol.*2012, **7**,490-493.
[11] Yokoyama T, *Phys.Rev.B* 2008, **77**, 073413.
[12] Haugen H, Huetas-Hernando D, Brataas A, *Phys.Rev.B* 2008, **77,** 115406.
[13] Yamamoto M, Shimazaki Y, Borzenets I V, Taracha S, *J.Phys. Soc. Jpn.* 2015,**84**, 121006.

IOP Conf. Series: Journal of Physics: Conf. Series **827** (2017) 012012 doi:10.1088/1742-6596/827/1/012012

Spin-polarized electron transport in hybrid graphene-BN nanoribbons

Song Gao[1],Wei Lu[1],Guo-Hui Zheng[1], Yalei Jia[1], and San-Huang Ke[1,2,*]

[1]MOE Key Labortoray of Microstructured Materials, School of Physics Science and Engineering, Tonji University, 1239 Siping Road, Shanghai 200092, P.R. China

[2]Beijing Computational Science Research Center, 10 Dongbeiwang West Road, Beijing 100094, P.R. China

E-mail: *shke@tongji.edu.cn

Abstract. The experimental realization of hybrid graphene and h-BN provides a new way to modify the electronic and transport properties of graphene-based materials. In this work, we investigate the spin-polarized electron transport in hybrid graphene-BN zigzag nanoribbons by performing first-principles nonequilibrium Green's function method calculations. A 100% spin-polarized electron transport in a large energy window around the Fermi level is found and this behavior is independent of the ribbon width as long as there contain 3 zigzag carbon chains. This behavior may be useful in making perfect spin filters.

1. Introduction

In recent years, graphene and monolayer hexagonal boron nitride (h-BN) have been attracted much attention for their potential applications in next-generation electronics[1, 2, 3, 4, 5] after they were successfully synthesized in experiments by different methods[6, 7, 8, 9]. Zigzag-edged graphene nanoribbons (zGNRs) exhibit spin-polarized edge states which are localized around the two edges, as have been observed in monatomic step edges of graphite by scanning probe techniques[11, 12]. While the zigzag-edged h-BN nanoribbons (zBNNRs) are found to be spin degenerated except for bare ones[13]. Experimentally, people have demonstrated the ability to make graphene-BN stacks[14] as well as domain hybridized layers[10]. Very recently, Sutter et al. studied the growth and interface formation of monolayer graphene-BN heterostructures on ruthenium substrates by using a combination of in situ microscopy techniques[15]. All these advances in hybridizing conducting graphene with insulating h-BN offer a new scheme to tune the electronic properties of these low-dimensional materials.

Over the past few years, the electronic and transport properties of GNRs modified with BN domains or BNNRs have been investigated extensively. For instance, people successfully tuned the band gap of GNRs by introducing BN domains or BNNRs[16, 17, 18]. Furthermore, several approaches have been proposed to control the spin transport in GNR/BNNR hybrid systems, which may be useful in spintronics. Dong et al. found that a BNNR seamlessly connected between two pieces of GNRs can result in 100% electron spin polarization[18]. He et al. found that the currents with different spin configurations display different behavior for hybridized zBNNR$_m$-zGNR$_n$ systems with certain n [16], but they did not find a way to achieve 100% spin polarization. Calculations by Dubois et al. showed that an erect heterojunction made of a bare zBNNR and zGNR can be a passive spin filter[19].

We note that highly spin-polarized electron transport may be produced by destroying the spin symmetry of zGNRs as in the case of parallel hybrid zGNR/zGNNR systems. In this work, we investigate

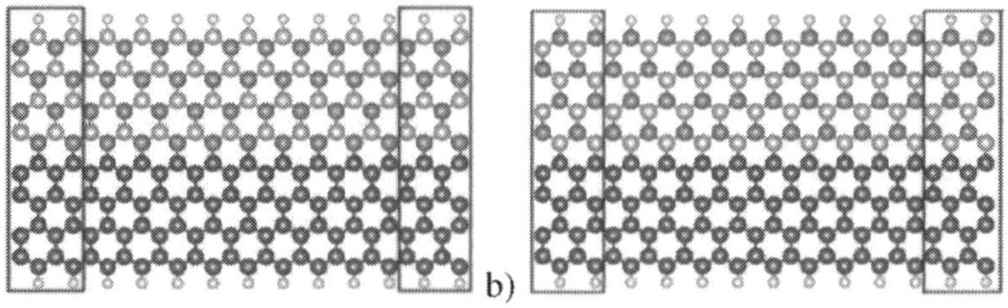

Figure 1. (Color online) Structures of hybrid systems of 4-zGNR/4-zBNNR. (a) C atoms are connected to B atoms, (b) C atoms are connected to N atoms. The brown, pink, gray, and green balls denote the C, H, N, and B atoms, respectively.

the electron transport properties of the parallel hybrid zGNR/zBNNR systems with different widths by performing first-principles calculations based on the nonequilibrium Green's function (NEGF) method combined with the density functional theory (DFT). Two different cases are considered: The zGNR is connected to the zBNNR via B and N atoms, respectively. It is found that highly spin-polarized electron transport in a large energy window around the Fermi level is presented. Especially, a 100% spin-polarized transport appears, which is independent of the ribbon width as long as there contain 3 zigzag carbon chains in the zGNR.

2. Models and Computation

We first consider a zGNR consisting of 8 zigzag carbon chains (8-zGNR). Then we gradually substitute these carbon chains with BN chains to construct parallel zGNR/zBNNR hybrid nanoribbon systems. The connection is made via the B atoms (system A) or N atoms (system B). The two systems are denoted by mBn and mNn, respectively, with m and n being the numbers of C chains and BN chains, respectively. The two edges of the hybrid nanoribbons are saturated by H atoms. As an example, the structures of the 4B4 and 4N4 nanoribbons are shown in Fig.1. There is a spin-polarized edge state localized around the edge of the zGNR and we choose its spin state as spin majority (α-spin) (This can always be realized by, for example, applying an external magnetic field[20]). To investigate the electron transport through the nanoribbon systems, we adopt the NEGF method combined with first-principles DFT calculations[21]. The periodic nanoribbon systems are divided artificially into three subsystems: left lead, scattering region, and right lead (see Fig.1). The self-consistent Kohn-Sham Hamiltonian of the scattering region and the self-energies of two semi-infinite leads are used to construct a single-particle Green's function from which the transmission coefficient at any energy is calculated. The computational techniques have been described in details previously[21]. For the DFT electronic structure calculations, we employ a double-ζ plus polarization (DZP) numerical basis set for all atomic species[22, 23]. The optimized Troullier-Martins pseudopotentials[24] are adopted to describe the atomic cores and the PBE version of the generalized gradient approximation (GGA)[25] is used for the electron exchange and correlation. The atomic structures of the systems studied are fully optimized by minimizing the forces on atoms to be smaller than 0.005 eV/Å.

3. Results and Analyses

To begin with, we present the transmission functions of pristine 8-zGNR with ferromagnetic (FM) and antiferromagnetic (AFM) spin configurations, as shown in Fig.2. For the FM configuration (see Fig.2(a)), the system is metallic and there is one α channel and one β channel around the Fermi energy. Away from the Fermi energy, there is an α plateau around -0.3eV and a β plateau around 0.25eV, which is consistent with the spin-polarized band structure shown in Fig.2(c). For the AFM configuration (see Fig.2(b)), the two spins are degererate and the system becomes semiconducting with a band gap of about 0.5eV, which is consistent with the spin-unpolarized band structure shown in Fig.2(d). These results agree well with a

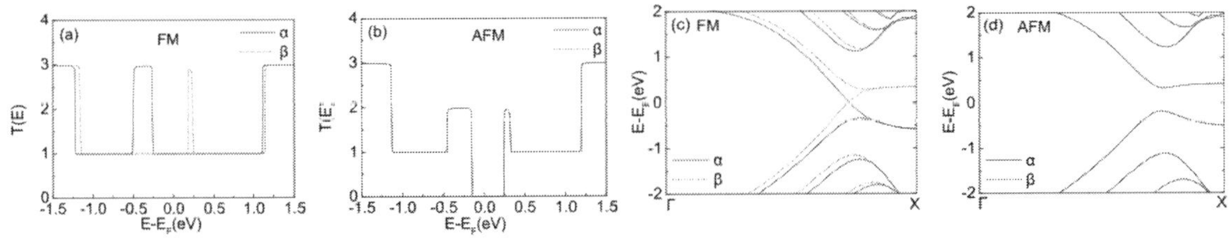

Figure 2. (Color online) Transmission functions of pristine 8-zGNR with (a) FM spin configuration and (b) AFM spin configuration. (c)(d) Band structures of 8-zGNR with FM and AFM configuration, respectively.

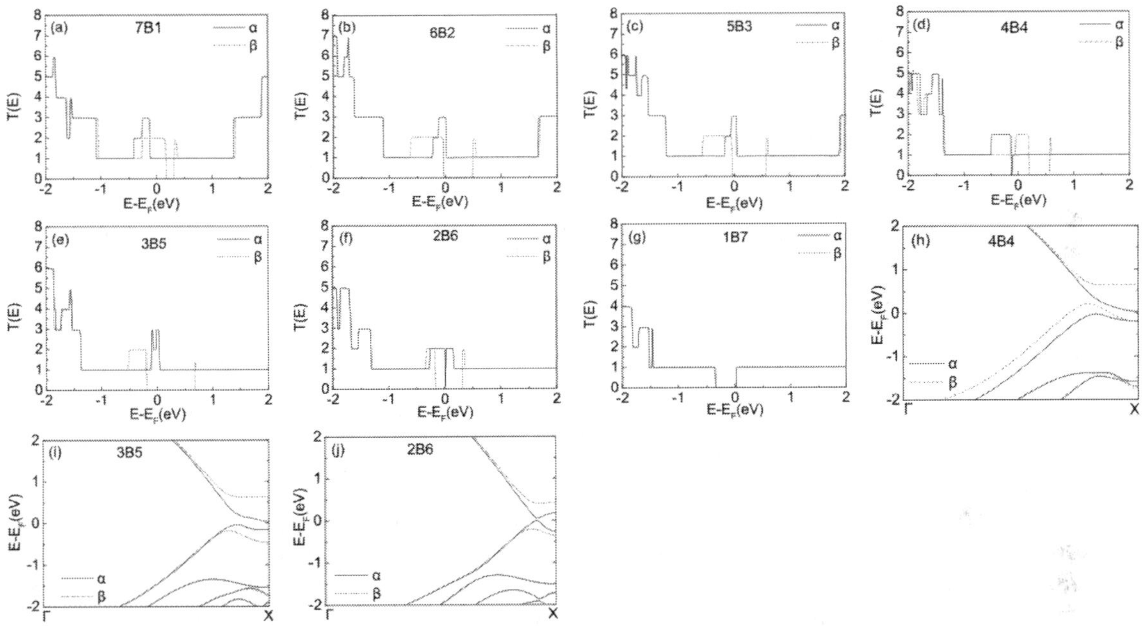

Figure 3. (Color online) (a)-(g) Transmission functions of different hybrid nanoribbons (from 7B1 to 1B7). (h)-(j) Band structures of 4B4, 3B5, and 2B6.

previous work about the transport in pristine 8-zGNRs[26].

Now, let us look at the transport properties of system A (connected via B atoms). The calculated transmission functions (T(E)) are given in Figs.3(a)-(g). The first thing to note is the significant spin polarization around the Fermi energy for all cases calculated. When the width of BN doesn't exceed half of the ribbon width (from 7B1 to 4B4), the α-spin shows the analogous behavior of pristine 8-zGNR with FM configuration while the β-spin shows the analogous behavior of pristine 8-zGNR with AFM configuration except for the fact that the Fermi level is shifted to lower energies for the two spins (see Figs.3 (a)-(d) vs Fig.2).

To understand the special transport properties, we calculate the band structures of these systems. Our calculation shows that their band structures (from 7B1 to 4B4) are similar, therefore, we take the band structure of 4B4 as an example to analyze (see Fig.3(h)). One can find that the α-spin band and β-spin band show indeed analogous behavior to those of the pristine 8-zGNR with FM and AFM spin configuration, respectively.

When the width of BN exceeds half of the ribbon width (from 3B5 to 1B7, see Fig.3(e)-(g)), the

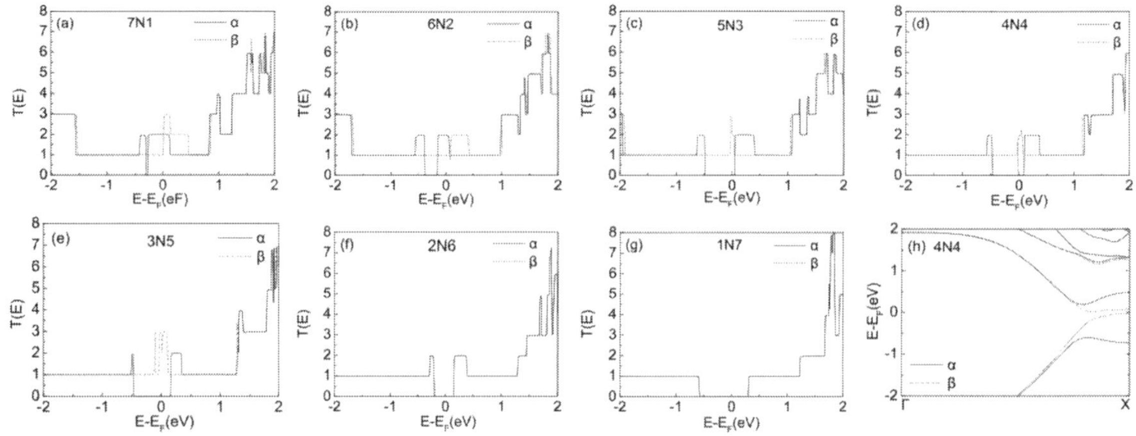

Figure 4. (Color online) (a)-(g) Transmission functions of different hybrid nanoribbons (from 7N1 to 1N7). (h) The band structure of 4N4.

Fermi level moves back to the transport gap for the β-spin. As a result, for 3B5 and 2B6 there comes a complete (100%) spin polarization in a large energy range around the Fermi level as shown in Fig.3 (e)-(f). This can also been seen clearly in the band structures of 3B5 and 2B6 shown in Fig.3 (i) and (j): The Fermi level is in the band gap of the β-spin, where the α-spin bands exist. As for 1B7, that is, there is only one zigzag carbon chain presented in the ribbon, the spin polarization disappears leaving a transmission gap around the Fermi level (see Fig.3 (g)).

Next, let us look at the transport behavior of system B (from 7N1 to 1N7) as shown in Fig.4. One can see that the behavior is quite different from system A: Now the α-spin shows the analogous behavior of pristine 8-zGNR with AFM configuration while the β-spin shows the analogous behavior of pristine 8-zGNR with FM configuration, just being opposite to system A. This can also be seen in the band structure of 4N4 shown in Fig.4 (h) for example: In contrary to 4B4, here the α-spin bands show the behavior of pristine 8-zGNR with AFM configuration while β-spin bands show the behavior of pristine 8-zGNR with FM configuration. In the case of 3N5, a 100% spin polarization appears within the energy window of (-0.50, 0.16)eV as shown in Fig.4 (e). In this energy window the α-spin channel is completely blocked while there come large transmission coefficients for the β-spin channel. As for the 2N6 and 1N7 systems, the spin polarization disappears (see Figs.4 (f) and (g)) leaving a transport gap of 0.3 and 0.9eV, respectively. This provides a reliable way to modifying largely the band gap of the hybrid nanoribbons.

As mentioned above, the transport properties of 3B5 and 3N5 is very interesting. Both of them can provide a large energy windows around the Fermi level where the electron transport is 100% spin polarized and, therefore, may be used to produce high-performance spin filters. To check whether or not this behavior is general for hybrid nanoribbons consisting of 3 zigzag carbon chains, we further calculate other systems with different ribbon widths: 3B1, 3B3, 3B7, and 3N1, 3N3, 3N7. The transmission functions are given in Fig.5. As it can be seen, in all the cases there exists the energy window of 100% spin polarization. Furthermore, the range of this energy window increases gradually with increasing ribbon width, indicating that this behavior is general for hybrid zGNR/zBNNR systems consisting of 3 zigzag carbon chains.

To understand this 100% spin polarization, we analyze the band structure and projected density of states (PDOS) of 3N5, as shown in Fig.6. There is a band gap for the α-spin energy bands which are just analogous with the band structure of pristine 8-zGNR with AFM configuration. However, for the β-spin the band gap disappears. The PDOS indicates that the two bands in the α-spin band gap come from the hybridization of p_z orbital of C atoms and p_z orbital of N atoms, indicating that the complete spin polarization effect stems from the interaction between the GNR and BNNR.

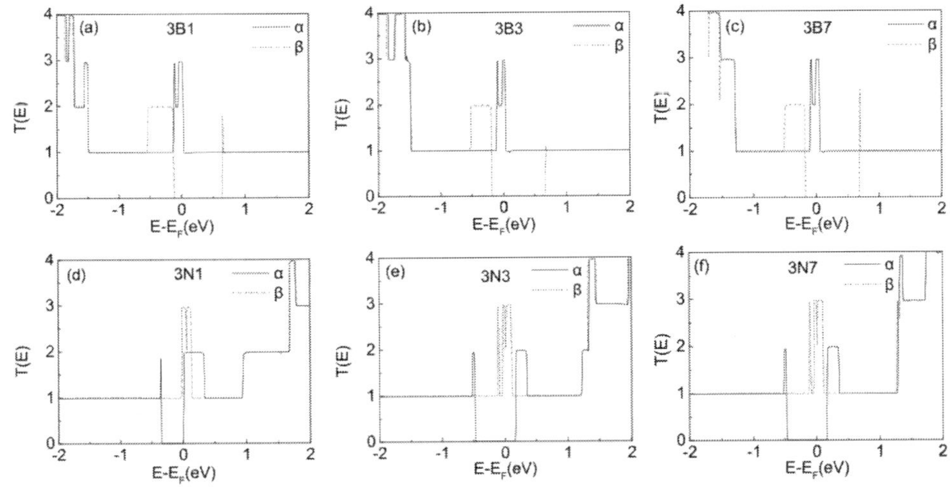

Figure 5. (Color online) Transmission functions of (a) 3B1, (b) 3B3, (c) 3B7, and (d) 3N1, (e) 3N3, (f) 3N7.

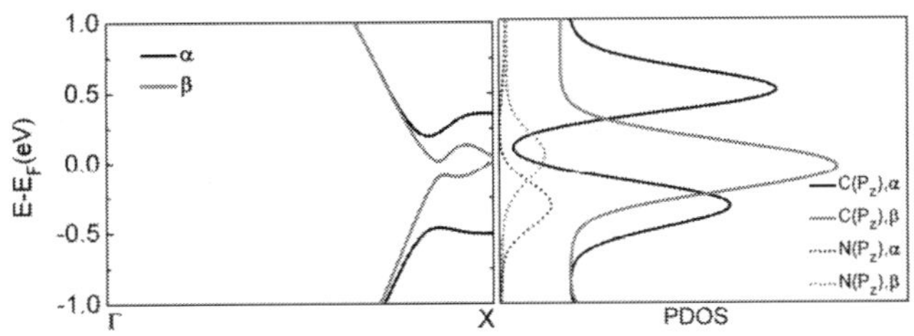

Figure 6. (Color online) The band structure and PDOS of 3N5.

4. Conclusions

By performing first-principles calculations based on the nonequilibrium Green's function method combined with density functional theory, we have studied the electron transport behavior in hybrid zGNR/zBNNR nanoribbons with different connection atoms and ribbon widths, which can be now fabricated in experiments. Significant spin-polarized transport behavior is found in most cases. Especially, hybrid nanoribbons consisting of 3 zigzag carbon chains will provide a large energy window around the Fermi level where the electron transport is 100% spin polarized. This is a general behavior regardless of the connection atoms (B or N atom) and the ribbon width. This behavior may be used in making perfect spin filters.

5. Acknowledgments

This work was supported by the National Natural Science Foundation of China under Grants No. 11174220 and 11374226.

References

[1] A. K. Geim, 1992 Science **324**, 1530.
[2] A. H. Castro Neto, F. Guinea, K. S. Novoselov and A. K. Geim, 2009 Rev. Mod. Phys**81**, 183.
[3] A. Nad, K. Raidongia, K. P. S. S. Hembram, R. Datta, U. V. Waghmare, and C. N. R. Rao, 2010 ACS Nano **4**, 1539.
[4] D. Golberg, Y. Bando, Y. Huang, T. Terao, C. Tang, and C. Zhi, 2010 ACS Nano **4**, 2979.
[5] M. L. Hu, Z. Z. Yu, K. W. Zhang, and L. Z. S. J. X. Zhong, 2011 J. Phys. Chem. C **115**, 8260.

[6] K. S. Novoselov, A. K. Geim, S. V. Morozov, D. Jiang, Y. Zhang, S. V. Dubonos, I. V. Grigorieva, and A. A. Firsov, 2004 Science **306**, 666.
[7] W. Q. Han, L. Wu, Y. Zhu, K. Watanabe, and T. Taniguchi, 2008 Appl. Phys. Lett. **93**, 223103.
[8] C. Jin, F. Lin, K. Suenaga, and S. Iijima, 2009 Phys. Rev. Lett. **102**, 195505.
[9] N. Alem, R. Erni, C.kisielowski, M. D. Rossell, W. Gannett, and A. Zettl, 2009 Phys. Rev. B **80**,155425.
[10] L. Ci, L. Song, C. Jin, D. Jariwala, D. Wu, Y. L. A. Srivastava, Z. F. Wang, K. Storr, and L. Balicas, 2010 Nat. Mater **9**, 430.
[11] Y. Kobayashi, K.-I. Fukui, T. Enoki, K. Kusakabe, and Y. Kaburagi, 2005 Phys. Rev. B **71**, 193406.
[12] Y. Niimi, T. Matsui, H. Kambara, K. Tagami, M. Tsukada, and H. Fukuyama, 2007 Phys. Rev. B **73**, 085421.
[13] M. Topsakal, E. Akturk, and S. Ciraci, 2009 Phys. Rev. B **79**, 115442.
[14] Z. Liu, L. Song, S. Zhao, J. Huang, J. Lou, and P. M. Ajayan, 2011 Nano Lett. **11**, 2032.
[15] P. Sutter, R. Cortes, J. Lahiri, and E. Sutter, 2012 Nano Lett. **12**, 4869.
[16] J. He, K. Q. Chen, Z. Q. Fan, L. M. Tang, and W. P. Hu, 2010 Appl. Phys. Lett. **97**, 193305.
[17] L. Song, L. Balicas, D. J. Mowbray, R. B. Capaz, K. Storr, L. Ci, D. J, and S. K, 2012 Phys. Rev. B **86**, 075429.
[18] J. C. Dong and H. Li, 2012 J. Phys. Chem. C **116**, 17259.
[19] S. M.-M. Dubois, X. Declerck, J.-C. Charlier, and M. C. Payne, 2013 ACS Nano. **115**, 10836.
[20] Y. Son, M. Cohen, and S. Louie, 2006 Nature **444**, 347.
[21] S.-H. Ke, H. Baranger, and W. T. Yang, 2004 Phys. Rev. B **70**, 085401.
[22] D. Sánchez-Portal, P. Ordejon, E. Artacho, and J. M. Soler, Int. 1997 J. Quantum Chem.**65**, 453.
[23] J. M. Soler, E. Artacho, J. D. Gale, A. Garcia, J. Junquera, P. Ordejón, and D. Sónchez-Portal, 2002 J. Phys.:Condens. Matter **14**, 2745.
[24] N. Troullier and J. L. Martins, 1991 Phys. Rev. B **43**, 1993.
[25] J. P. Perdew, K. Burke, and M. Ernzerhof, 1996 Phys. Rev. Lett. **77**, 3865.
[26] S. M.-M. Dubois, Z. Zanolli, X. Declerck, and J.-C. Charlier, 2009 Eur. Phys. J. B **72**, 1.

IOP Conf. Series: Journal of Physics: Conf. Series **827** (2017) 012013 doi:10.1088/1742-6596/827/1/012013

Pressure-induced metallic ferromagnetism in antiferromagnetic insulator LaMnAsO

HuaiBao Tang[a,*], Guang Li[a], Qun Yang[a], XueQing Zuo[a], HaiJun Zhang[a] and DaYong Liu[b,]**

a School of Physics and Materials Science, Anhui University, Hefei 230039, China

b Key Laboratory of Materials Physics, Institute of Solid State Physics, Chinese Academy of Sciences, P. O. Box 1129, Hefei, Anhui 230031, China

E-mail: * `hbtang@ahu.edu.cn`, ** `dyliu@theory.issp.ac.cn`

Abstract. Based on the first-principles calculations, we investigate the effects of the hydrostatic pressure on the layered oxypnictide LaMnAsO, and predict that it undergoes a transition from a G-type antiferromagnetic (AFM) insulator to a ferromagnetic (FM) metal at the critical pressure $P_c \sim 18.0$ GPa, accompanied by a high-spin state with $m_{Mn} = 2.65\ \mu_B$ to a low-spin one with $m_{Mn} = 1.63\ \mu_B$ transition, or magnetic moment collapse. Our calculations identify that it is the hydrostatic pressure that collapses the G-type AFM gap, and the large density of state (DOS) around Fermi level stabilizes the FM metallic phase due to the Stoner instability. The pressure-induced AFM insulator to FM metal transition opens up the possibility of two state device applications in the oxypnictide LaMnAsO compound.

1. Introduction

Since the discovery of hight-temperature superconductivity in fluorine-doped LaFeAsO [1, 2, 3], the quaternary compounds with ZrCuSiAs-type (abbreviated as 1111-type) structure have recently attracted considerable interest as advanced electronic functional materials. With full replacement of the Fe sites by other transition metals, a larger number of layered transition metal arsenide oxide *LnM*AsO (*Ln*: lanthanoid, *M*: transition metal) have been synthesized and studied experimentally, meanwhile they exhibit many interesting properties: besides LaFeAsO serving as mother compound for hight-temperature superconductor, they exhibit giant magnetoresistance and rare earth magnetism in *Ln*MnAsO [4, 5, 6, 7], itinerant ferromagnetism in *Ln*CoAsO [8, 9, 10, 11], low temperature superconductivity and Kondo physics in *Ln*NiAsO [12, 13, 14], and glassy magnetism in *Ln*RuAsO [15].

Of particular interest is the Mn-based 1111 material LaMnAsO, which is an in-plane AFM insulator with an experimental gap of $\sim$ 1.1 eV [5,16]. The Mn atoms order in a checkerboard fashion termed as G-type antiferromagnetism, with the spins aligned collinear to the c axis [17]. The Néel temperature is 360 K, and the local moment at the Mn site is large, $\sim 3.34\ \mu_B$ [17], contrary to the small moment of $0.4 \sim 0.8\ \mu_B$ per Fe atom seen in the stripe-like (C-type) AFM ground state of the isostructure LaFeAsO [18]. Band structure calculations [19] indicate as well that LaMnAsO is G-type AFM despite an even smaller gap values of $\sim$ 0.2 eV.

The AFM and insulating ground state of LaMnAsO could recall those of the parent compound of cuprate superconductors, where the highest T_c occurs when pressure or doping transform the localized electrons in the AFM insulator into itinerant carriers in the metal. With those in view,

IOP Conf. Series: Journal of Physics: Conf. Series **827** (2017) 012013 doi:10.1088/1742-6596/827/1/012013

even larger T_c may be desirable in the isostructural Mn analog of Fe-based superconductors, provided that a stabilizing metallic state is achieved. Indeed, the essentially metallic behavior was displayed in $La_{1-x}Sr_xMnAsO$ via hole doping [20], and associated with metallization via electron doping, ferromagnetism with negative magnetoresistance was also observed in $LaMnAsO_{1-x}H_x$ for $x > 0.14$ [21]. It is natural to ask whether or not the AFM insulator to FM metal transition occurs in the pressured LaMnAsO.

In this paper we present the magnetic properties of the layered manganese oxyarsenide LaMnAsO under hydrostatic pressure obtaining by the first-principles calculations. Our results clearly show that a transition from a low-pressure AFM insulator to a high-pressure bulk FM metal when the pressure is increased to the critical value $P_c \sim 18$ GPa. Such a magnetic property could be useful in two state sensor device. The rest of the paper is organized as follows: we describe the crystal structures of pressured LaMnAsO and computational method in Sec. II, and present the numerical results and discussions in Sec.III, and Sec.IV is devoted to the summary.

2. Structure and computational methods

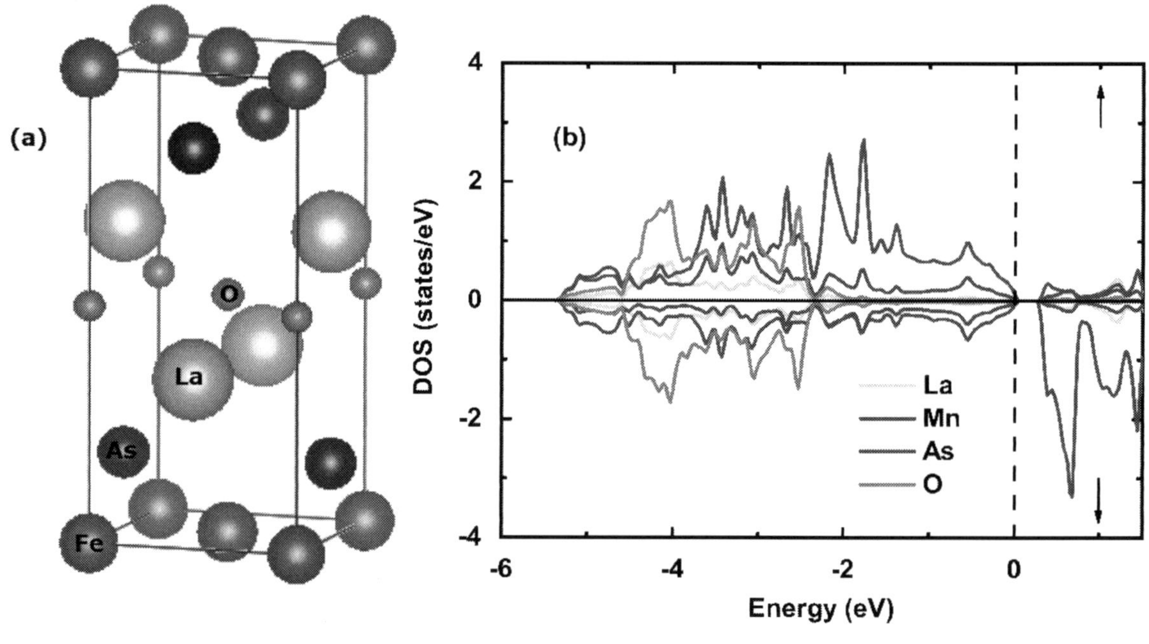

Figure 1. (Color online) (a) Unit cell of crystal structure of LaMnAsO and (b) Spin-polarized atom-projected density of states (PDOS) of the G-type AFM state in ambient pressure obtained by GGA.

As shown in Fig. 1 (a), LaMnAsO has a layered tetragonal crystal structure with the space group $P4/nmm$, where the charge donor LaO layers are sandwiched between the functional MnAs layers in an ABA sequence. In the essential MnAs layers, each Mn cation is caged by an edge-sharing As tetrahedron, defining a two-dimensional square Mn atomic lattice. To study the electronic structure and magnetic properties of LaMnAsO under the hydrostatic pressure, we have used the following structural optimization procedure: we fix the unit cell volume and optimize the c/a ratio and all internal atomic coordinates for the tetragonal lattice. Manually changing the lattice volume allows for the investigation of different hydrostatic pressure, and the Birch-Murnaghan equation of state is used to estimate the theoretical pressure.

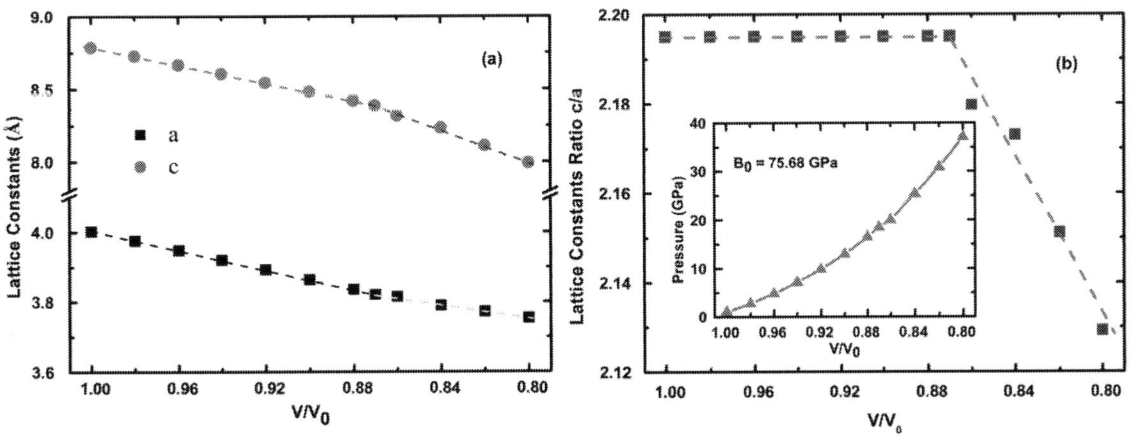

Figure 2. (Color online) (a) The tetragonal lattice cell parameters of LaMnAsO, a and c, versus unit cell volume. (b) The c/a ratio versus unit cell volume, inset: the pressure intensity versus unit cell volume with the bulk modulus B_0.

The calculations reported here are performed by using full-potential linearized augmented plane wave (FP-LAPW) method based on density functional theory (DFT), as implemented in the WIEN2k package [22]. For the exchange correlation potential, we used the generalized gradient approximation (GGA) in the Perdew-Burke-Ernzerhof variant [23]. The number of k points was 1000 for the tetragonal unit cell. The C-type AFM unit cell size is doubled to $\sqrt{2} \times \sqrt{2} \times 1$. The self-consistent calculations are considered to be converged when the total energy difference of the crystal does not exceed 0.1 mRy and that in the total electronic charge does not exceed 10^{-4} electronic charge at consecutive steps. All independent internal atomic coordinates are optimized until the corresponding forces are less than 1 mRy/$a.u.$.

3. Results and discussions

Table 1. The GGA results of LaMnAsO at ambient pressure. The NM state is the referent state for the energy.

Magnetic Order	Energy (meV/Mn)	Moment (μ_B/Mn)	Gap (eV)
NM	0.0	0.0	-
FM	-289.06	1.89	-
G-AFM	-544.07	3.21	0.31
C-AFM	-318.77	3.17	0.24

At ambient pressure, the optimized lattice constants for LaMnAsO are $a = 4.003$ Å and $c = 8.816$ Å, which is about 2.6% smaller than the experimental value [16, 17]. Using the optimized lattice constants and the numerical internal coordinates, several magnetic orders have been calculated to determine the ground state. The numerical results are summarized in Table. 1. In excellent agreement with the neutron scattering experiments [5, 16], the G-type AFM state has the lowest energy. At the same time, the experimental local moments of $\sim 3.34\ \mu_B$ [17] could be well captured by the present GGA calculations, in which $m_{Mn} \sim 3.21\ \mu_B$. The notable reduced moment suggests an itinerant character and strong spin fluctuations of the AFM ground state, and could be regarded an a primary factor of metallization by the external pressure. The spin-polarized atom-projected density of states of the G-type AFM state is also plotted in Fig.

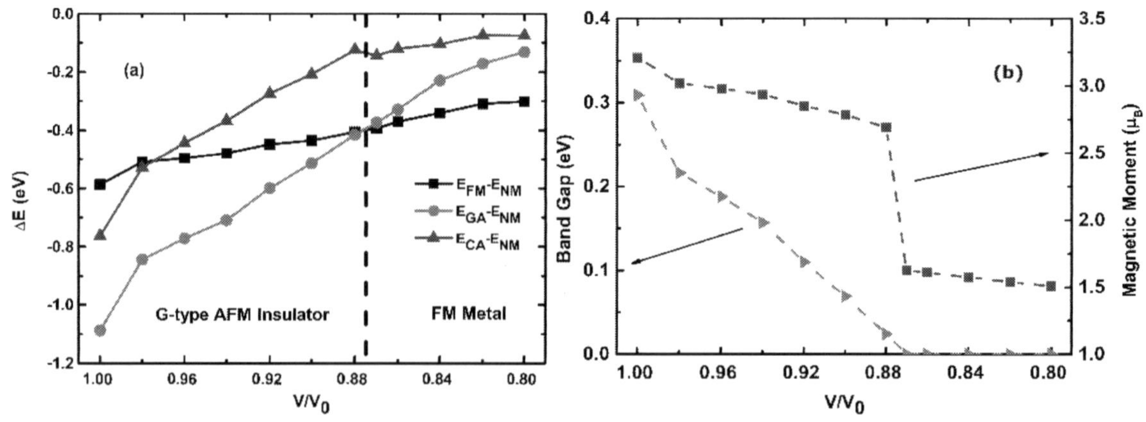

Figure 3. (Color online) (a) Total energy differences of FM (black line), G-type AFM (red line) and C-type AFM (blue line) states with respect to NM state over the volume ratio form 1.0 to 0.8 GPa. (b) The GGA band gap and magnetic moment are plotted as functions of volume ratio.

1 (b). While an intrinsic insulating behaviour is clearly shown by a band gap of $\sim$ 0.31 eV at the Fermi level, another important feature of LaMnAsO compound is the relatively high PDOS near the bottom of conducting band, mostly from the Mn's $3d$ orbital. This suggests a possibility of a Stoner ferromagnetic instability against G-type AFM state provided that the G-type AFM gap is closed by the external pressure.

Using the above structural optimization procedure, we firstly show that LaMnAsO holds the expected tetragonal structure over the range of the volume ratio $\nu = V/V_0$ from 1.0 to 0.8, where V_0 is the volume at ambient pressure. The calculated variation of the a and c lattice parameters and the ratio c/a are shown in Fig. 2 (a) and (b), respectively. While the lattice parameters a and c decrease monotonically with the decrease in volume, there exits a distinct change of lattice parameters close to the ratio $V/V_0 \sim 0.87$, accompanied by a significant reduction in the ratio c/a. This means that the lattice parameter a becomes less compressible at high pressures, whereas the compressibility of the lattice parameter c shows a slight increase. The pressure intensity of different volume was quantified by using Birch-Murnaghan equation of state, and the numerical result is shown in the insert of Fig. 2 (b).

To study the relative stability of G-type AFM state against other possible magnetic state as increasing pressure, we perform numerical simulations on the three possible spin configurations (G-type AFM, C-type AFM, and FM) and the nonmagnetic (NM) state. Fig. 3 (a) displays the stabilization energies of these possible spin configurations relative to the NM state as a function of V/V_0. Obviously, the G-AFM phase exhibits the lowest energy under the low pressure, but the total energy difference between the G-type AFM and FM state decreases with the decrease in volume. At the critical value of $V/V_0 \sim 0.87$, amounting to a pressure value $P_c \sim 18$ GPa, a magnetic transition from the G-type AFM to FM state occurs, and additional decrease in volume further stabilizes the FM phase. The relationships among the volume V/V_0, band gap and local Mn moment are shown in Fig. 3 (b). When the volume reaches to the critical value $V_c \sim 0.87V_0$, the bulk energy gap is closed, indicating that an insulator-to-metal (I-M) conversion happens along with the G-type AFM to FM transition occurring simultaneously. It is worthwhile to point out that the calculated V_c (or P_c) value likely represents a lower bound of the volume (or pressure) where the phase transitions occur, since the GGA method usually underestimate the band gap. In addition, while the substantial spin polarization of the G-type AFM state could be diminished by the decrease in volume, a high-spin to low-spin moment collapse rightly when

IOP Conf. Series: Journal of Physics: Conf. Series **827** (2017) 012013 doi:10.1088/1742-6596/827/1/012013

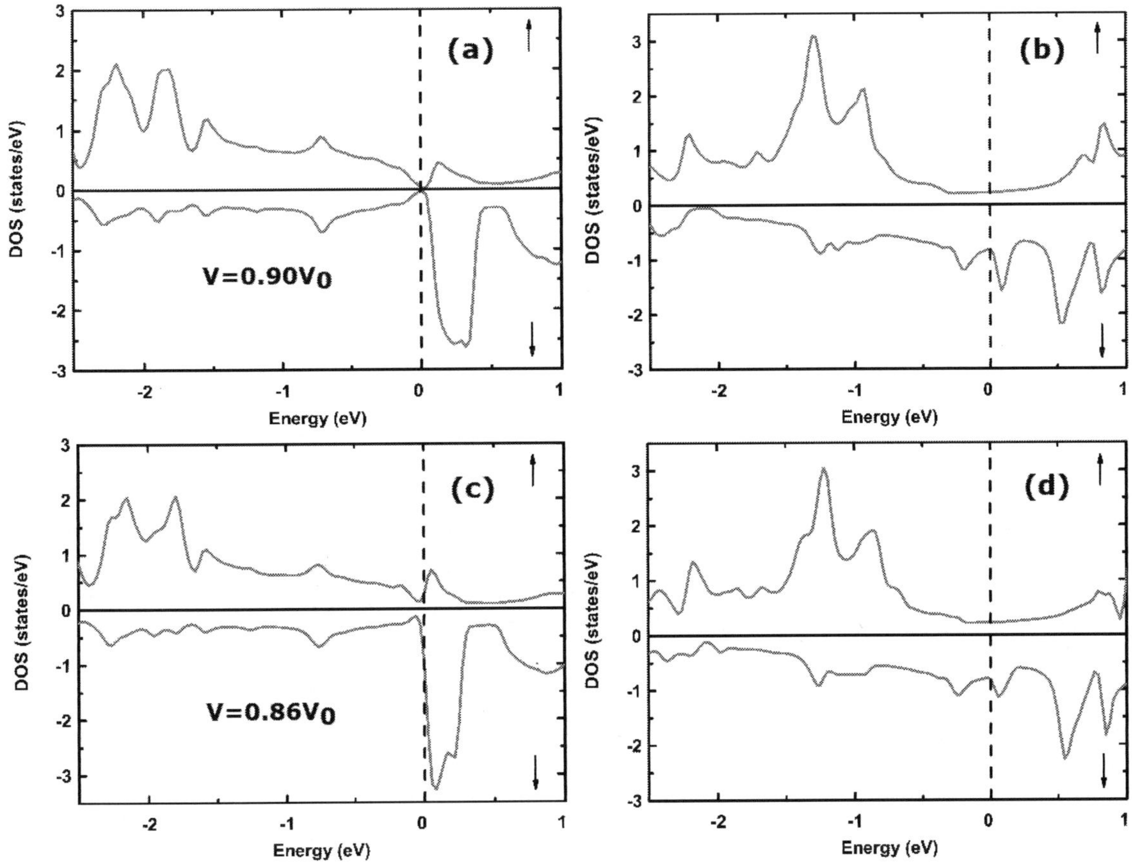

Figure 4. (Color online) Spin-polarized DOS of Mn at $V/V_0 = 0.9$ (a, b) and at $V/V_0 = 0.86$ (c, d) for G-type AFM (a, c) and FM (b, d) states, respectively.

the G-type AFM to FM transition comes at the critical V_c, as seen in Fig. 3 (b). Physically that is correlated with the Hund's metal nature of the FM state, in which the itineration of electrons enforces the spin up and down symmetrical distribution of electrons within the Mn 3d shell, but the Hund's rule coupling is unnecessarily reduced to zero under the present pressure, yielding a rapid reduction of the local moment down to the finite value $\sim 1.63\ \mu_B$.

To further illustrate the profound magnetic and electronic transition of the pressured LaMnAsO, the spin-polarized density of states of Mn cation are plotted in Fig. 4 for both of the G-type AFM and FM states at $V/V_0 = 0.90$ and $V/V_0 = 0.86$. Despite a little increase of the density of states near the bottom of conducting band, one could clearly notice that the overall profile of density of states for both of the G-type AFM and FM cases do not change much as was previously appreciated. When the external pressure is higher than the critical value P_c, the band gap of the G-type AFM state has been closed and the Fermi level is shifted to the high density of states of the conducting band. Hence the Stoner instability make the system metallic FM ordering according to optimizing the kinetic energy of the polarized state and suppressing the density of states of the G-type AFM state at the Fermi level.

4. Summary

In summary, we have investigated the effect of hydrostatic pressure on the G-type AFM insulator LaMnAsO using a full potential linearized augmented plane-wave method. The I-M transition accompanied by the emergence of FM state is predicted at the critical pressure $P_c \sim 18.0$ GPa. Meanwhile a high-spin to low-spin moment transition is also revealed when the I-M transition occurs. It is the hydrostatic pressure that drives the collapse of the G-type AFM gap of LaMnAsO, and then the high density of state around Fermi level stabilizes the metallic FM phase due to the Stoner instability. Since the pressure modulation drives the insulator-metal transition, these results open up the possibility of two-state device applications based the oxypnictide LaMnAsO compound.

5. Acknowledgments

The author (H.B.) would like to thank LiangJian Zou for valuable suggestions, discussions and carefully reading of the manuscript. The work was financially supported by the National Sciences Foundation of China under Grant Nos. 11574315 and 21403001.

References

[1] Kamihara Y, Watanabe T, Hirano M, and Hosono H 2008 *J. Am. Chem. Soc.* **130** 3296
[2] Takahashi H, Igawa K, Arii K, Kamihara Y, Hirano M, and Hosono H 2008 *Nature* (London) **453** 3296
[3] Chen X H, Wu T, Wu G, Liu R H, Chen H, and Feng D L 2008 *Nature* (London) **453** 761
[4] Marcinkova A, Hansen T C, Curfs C, Margadonna S, and Bos J W G 2010 *Phys. Rev. B* **82** 174438
[5] Emery N, Wildman E J, Skakle J M S, Giriat G, Smith, R I and Mclaughlin A C 2010 *Chem. Commun.* **46** 6777
[6] Tsukamoto Y, Okamoto Y, Matsuhira K, Whangbo M H, and Hiroi Z 2011 *J. Phys. Soc. Jpn.* **80** 094708
[7] Wildman E J, Skakle J M S, Emery N, and Mclaughlin A C 2012 *J. Am. Chem. Soc.* **134** 8766
[8] Yanagi H, Kawamura R, Kamiya T, Kamihara Y, Hirano M, Nakamura T, Osawa H, and Hosono H 2008 *Phys. Rev. B* **77** 224431
[9] Ohta H and Yoshimura K 2009 *Phys. Rev. B* **80** 18409
[10] McGuire M A, Gout D J, Garlea V O, Sefat A S, Sales B C, and Mandrus D 2010 *Phys. Rev. B* **81** 104405
[11] Marcinkova A, Grist D A M, Margiolaki I, Hansen T C, Margadonna S , and Bos J W G 2010 *Phys. Rev. B* **81** 065511
[12] Li Z, Chen G, Dong J, Li G, Hu W, Wu D, Su S, Zheng P, Xiang T, Wang N, and Luo J 2008 *Phys. Rev. B* **78** 060505(R)
[13] Matsuishi S, Nakamura A, Muraba Y, and Hosono H 2012 *Supercond. Sci. Technol.* **25** 084017
[14] Luo Y, Han H, Tan H, Lin X, Li Y, Jiang S, Feng C, Dai J, Cao G, Xu Z, and Li S 2011 *J. Phys.: Condens. Matter* **23** 175701
[15] McGuire M A, Garlea V O, May A F, and Sales B C 2014 *Phys. Rev. B* **90** 014425
[16] Kayanuma M A, Hiramatsu V O, Kamiya A F, Hirano A F and Hosono B C 2009 *J. Appl. Phys.* **105** 073903
[17] Emery N, Wildman E J, Skakle J M S, Mclaughlin A C, Smith R I, and Fitch A N 2011 *Phys. Rev. B* **83** 14429
[18] Cruz C, Huang Q, Lynn J M, Li J Y, Ratcliff W, Zarestky J L, Mook H A, Chen G F, Luo J L, Wang N L, and Dai P C 2008 *Nature* **453** 899
[19] Xu G, Ming W, Yao Y, Dai X, Zhang S C, and Fang Z 2008 *Europhys. Lett.* **82** 67002
[20] Sun Y, Bao J, Feng C, Xu Z, and Cao G 2012 *Europhys. Lett.* **98** 17009
[21] Hanna T, Matsuishi S, Kodama K, Otomo T, Shamoto S, and Hosono H 2013 *Phys. Rev. B* **87** 020401(R)
[22] Blaha P, Schwarz K, Madsen G K H, Kvasnicka D, and Luitz J, 2001 *WIEN2K, An Augmented Plane Wave+ Local Orbitals Program for Calculating Crystal Properties* (Karlheinz Schwarz, Technische Universitat Wien, Austria)
[23] Perdew J, Burke K, and Ernzerhof M 1996 *Phys. Rev. Lett.* **77** 3865

Magnetization and specific heat studies of the low temperature anomalies in terbium iron garnet $Tb_3Fe_5O_{12}$

M Lahoubi[1], **W Wang**[2] **and V Varazashvili**[3]

[1] Badji Mokhtar-Annaba University, Physics Department, Laboratory L.P.S, Annaba, 23000 Algeria
[2] Beijing University of Chemical Technology, School of Science, Beijing 100029, China
[3] Ivane Javakhishvili Tbilisi State University, Institute of Inorganic Chemistry and Electrochemistry, Tbilisi 0179, Georgia

E-mail: mlahoubi@hotmail.fr, wangwei@mail.buct.edu.cn

Abstract. The macroscopic magnetic properties of the terbium iron garnet ($Tb_3Fe_5O_{12}$) single crystals were studied at low temperature by magnetizations $M_T(H)$ in magnetic fields up to 80 kOe and applied along the easy axis <111>. Maxima are observed near ~ 65 and ~ 47 K, respectively in *the curves of the* initial magnetic susceptibility $\chi_0(T)$, and the parameter $|b|(T)$ of the bH^2 term are associated with the non linear variation of the $M_T(H)$ magnetizations. From the specific heat $C_P(T)$ measurements, a great peak is observed near 57 K in the excess contribution of the Tb^{3+} ions $C_p^{ex}(T)$ to $C_P(T)$ which vanishes in the 170–208 K temperature range. The results are discussed with respect to the simultaneous signs of the "low-temperature point" T_B predicted by Belov at 58 K and the 'Schottky anomaly' detected experimentally by the presence in the $C_P^{ex}(T)$ curve of a $1/T^2$ term for $T > T_B$.

1. Introduction

The rare earth iron garnets $RE_3Fe_5O_{12}$ or REIGs (where RE^{3+} is a trivalent rare earth ion or the yttrium) or their mixed compounds are distinguished today by new fundamental interests [1], [2]. This is the case of $Tb_3Fe_5O_{12}$ or TbIG who presents large magnetodielectric effects in low external magnetic fields (H_{ex} < 0.2 T) at 2 K [3]. Furthermore, a significant coupling between the magnetic exchange and ligand-field excitations near 60–80 K with two distinct behaviours above and below 150 K were also reported [4]. It was shown recently [5], [6] that despite the <111> direction is the easy axis (EA) of magnetization at all temperatures, the magnetic structure changes from collinear to a 'double-umbrella' structure below 160 K. This change, well interpreted on basis of group theory analysis [7], is followed by an anomalous magnetic behaviour around 54 K. In this paper, one expects that this low-temperature anomalous behavior to be reflected with anomalies on some pertinent parameters, such as magnetic susceptibility, which may lead to the proximity of the so-called 'low-temperature point' of Belov T_B predicted at 58 K for TbIG [8 and refs. therein]. But we wish to focus on another important effect arising from the 'Schottky effect' which is able to give rise to anomaly in the magnetic specific heat contribution due to Tb^{3+} ions only in the in the region of T_B.

2. Experimental details and results

The as-prepared TbIG samples are two spherical single crystals grown by the standard flux method. The first one (sample 1) (diameter: 5 mm; weight: 427.00 mg) is the same that used recently for the magnetization measurements $M_T(H)$ in high DC magnetic fields H_{ex} up to 200 kOe [9]. Whereas the second one (sample 2), it consists of a small sphere (diameter: 3 mm; weight 197.88 mg). Using a conventional magnetometer at the Institute Néel CNRS Grenoble, France, the $M_T(H)$ curves were measured from 4.2 K to 300 K with external magnetic field H_{ex} up to 80 kOe and applied along the EA <111>. The curves $M_T(H)$ versus the internal magnetic field H are overall reported in $\mu_B/2Tb_3Fe_5O_{12}$. The compensation temperature T_{comp} at which M^{total}(TbIG) vanishes is equal to 243.5 K [9] and 249 K for samples 1 and 2, respectively. The specific heat $C_p(T)$ measurements were performed on both samples using an AC calorimeter based on the compensation of power method from 5 K to 300 K. The measuring principle is based on the differential $C_p(T)$ between the sample and a reference. The increase of temperature (T) can be made according to variable speeds. Several runs with a rising of T with a speed of 1 K/mn were applied and a run was made with a more precision between 4.2 K and 100 K at a slower speed of 0.5 K/mn. Herein, the two best reproducible runs labeled as "data 1" and "data 2" in J/K mol (one mole = 2TbIG) for samples 1 and 2, respectively, are reported.

T*ypical* $M_T(H)$ curves obtained for the sample 1 at different values of T are given in Fig. 1. At 4.2 K, the sample is apparently saturated at 3.5 kOe, whereas the saturation is not attained for $T > 4.2$ K.

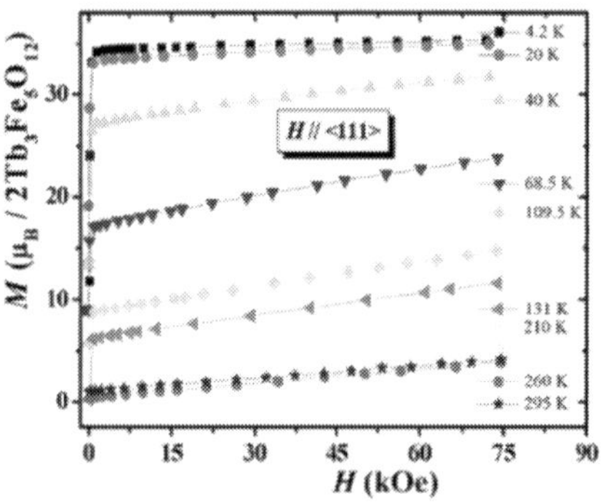

Figure 1. $M_T(H)$ vs H in the 4.2–300 K range.

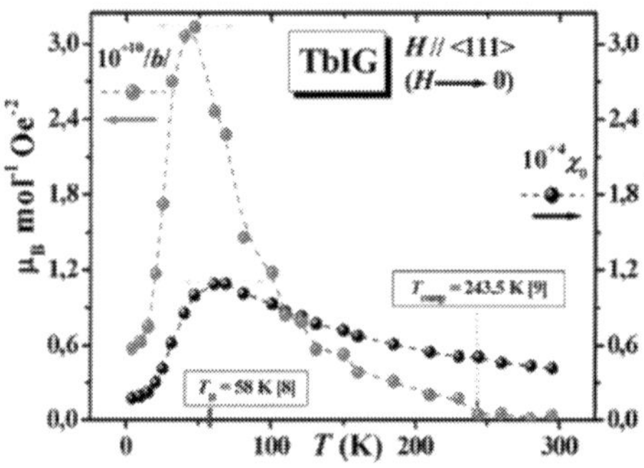

Figure 2. $10^{+4}\chi_0(T)$ and $10^{+10}|b|(T)$ vs T.

A non linear variation is shown with the exception of the T-region (220–260 K) below and above T_{comp} where the expected canted phases occur [9]. Both collinear and coaxial phases can be described using the previous expansion $M_T(H) = M_S(T) + \chi_0(T)H + b(T)H^2$ [9]. $M_S(T)$ identified with the spontaneous magnetization is equal to 34.35 μ_B/mol at 4.2 K. The initial magnetic susceptibility $\chi_0(T)$ is obtained by extrapolation at $H = 0$, and reaches the lowest value of 1.74×10^{-2} μ_B mol^{-1} kOe^{-1} at 4.2 K. Take into account the slight curvature of the curves, the parameter $b(T)$ is obtained. It is well-known that super-exchange interaction of Fe^{3+} in octahedral [a] and Fe^{3+} in tetrahedral (d) sites are almost identical for all REIGs [8]. Thus, we can determine the *spontaneous magnetization* M_{Tb} of the Tb^{3+} ions in dodecahedral {24c} site as $M_{Tb} \approx M^{total}$(TbIG) – M(YIG) in the basis of the two-sublattice model with the resultant of iron sublattices magnetization $M_{Fe} \approx M$(YIG). *At 4.2 K, the existence of a great difference in* M_{Tb} *between that (44.35* μ_B mol^{-1}) *of the coaxial structure and the calculated value in the hypothesis of both the collinear structure and the free ion value (54* μ_B mol^{-1}) *clearly reflects the effect of the* "double-umbrella" structure *of the* Tb^{3+} *moments, which is in agreement with the NPD results [5-7]. The plots of* $10^{+4}\chi_0(T)$ and the absolute values of b parameter $10^{+10}|b|(T)$ *are shown in Fig. 2.* Both curves increase when T decreases and show a maximum near ~ 65 K and ~ 47 K, respectively. *The measured* C_p(T) *data 1 and 2 plotted in the inset of Fig. 3 show a good agreement with those reported previously on powder sample for* T < *25 K [10]. One can see a qualitative agreement from 20 K to 200 K with a few discrepancy of about 6% in the* T-*range 200–298.15 K, as well as for the data 2, with the results obtained on powder sample prepared by the standard high-*T *solid-state* *reaction method [11]. Above 25 K, the $C_p(T)$ data 1 are in good agreement with those measured on a sample prepared by a sol-gel powder method [12].*

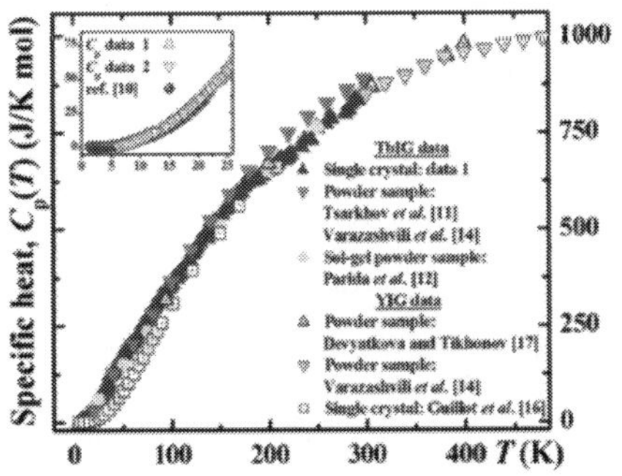

Figure 3. Plots of C_p data 1 with those found in the literature for TbIG and YIG. In inset, C_p data 1 and 2 are compared those in [10].

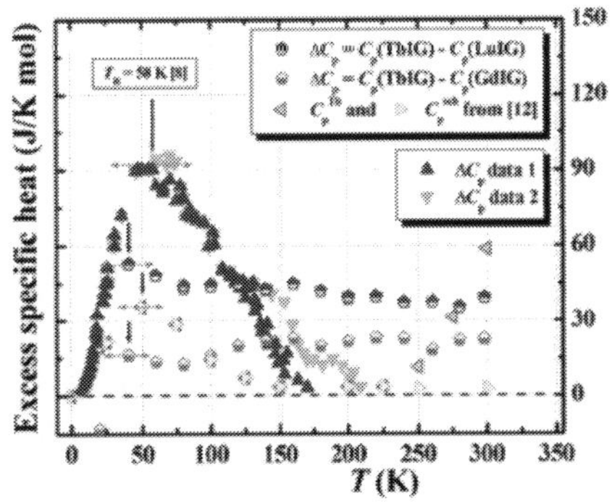

Figure 4. Plots of ΔC_p data 1 and 2 compared with other determinations of ΔC_p and with the calculated C_p^{Tb} and C_p^{sch} contributions [12].

3. Phenomenological analysis of the specific heat and discussion

As for all REIGs, C_p^{total}(TbIG) can be written simply as follows

$$C_p^{total}(\mathrm{TbIG}) = C_p^{nuc} + C_p^{reg} + C_p^{mag} + C_p^{sch} \tag{1}$$

where C_p^{nuc} arises from the interactions between nuclear and electron spins, C_p^{reg} named as "regular part" in [13], [14] represents both lattice and dilatation contributions, C_p^{magn} is the cooperative part associated with ferrimagnetic disordering process, C_p^{sch} is the magnetic non-cooperative electronic Schottky contribution caused by redistribution of 4f-electrons in the Tb^{3+} ions over the excited levels during the increasing of T. Define C_p^{MFe} and C_p^{Tb} are related to the contributions of M_{Fe} and M_{Tb} sublattice magnetization, and then, the magnetic contribution C_p^{magn} resulting from the competition between three types of anisotropies (spin-orbit coupling, crystalline field and super-exchange interactions) is simply described as $C_p^{magn} = C_p^{MFe} + C_p^{Tb}$. Then, C_p^{nuc} can be neglected above 6 K for TbIG [10], C_p^{total}(TbIG) can be expressed as follows

$$C_p^{total}(\mathrm{TbIG}) = C_p^{reg} + C_p^{MFe} + C_p^{Tb} + C_p^{sch} \tag{2}$$

So, on the basis of the Debye model, the evaluation of C_p^{reg} cannot be simply obtained [15]. We assume that TbIG has the same value of C_p^{reg} with that of YIG. Besides this, the resultant M_{Fe} is saturated below T_{comp} for TbIG [9]. Meanwhile, M_{Fe} can not affect C_p^{total}(TbIG). Therefore, we can put $C_p^{reg} + C_p^{MFe} \approx C_p$(YIG) in equation (2). The expected excess magnetic contribution C_p^{ex} can be written as follows

$$C_p^{ex} = \Delta C_p \approx C_p^{total}(\mathrm{TbIG}) - C_p(\mathrm{YIG}) = C_p^{Tb} + C_p^{sch} \tag{3}$$

It must be noted that, in C_p(YIG), neither the 'low-T' of Belov T_B nor 'Schottky effect' are expected. The reason is that the ground state of the Y^{3+} ion is in a S-state, but it is not the case for the magnetic Tb^{3+} ion. In Fig 3, we have plotted the C_p(YIG) data measured on single crystal from 5.3 K to 208.6 K [16] with those reported for YIG powder samples [14], [17] in order to smooth more data in the large T-range. It is noted that the Debye temperature θ_D and lattice parameters vary negligibly through the REIGs [18]. Since θ_D is proportional to $(\bar{m})^{-1/2}$, where $\bar{m}$ is an average mass of the atoms in the unit cell, and θ_D(TbIG) is estimated to be larger than θ_D(YIG) by about 15% if $\bar{m}$ is assumed proportional to the molecular weight. A large excess in the C_p^{ex} curve is observed in Fig. 4 with a broad peak reaching ~ 93 J/K mol nearby 57 K for both data 1 and 2. After a rapid decreasing with an increasing of T, C_p^{ex} disappears near 170 and 208 K, respectively. After subtracting those of LuIG which was considered as an isostructural baseline [13], [14], the C_p^{ex} data deduced by the C_p^{total}(TbIG) data are also shown in Fig. 4, where a maximum (about ~ 53 J/K mol) is observed at near 45 K. The differences between maxima and low-T-anomalies possibly arise from the large difference between atomic mass of Y and Lu elements. Then, one can estimate that θ_D(YIG) should be 15% larger than θ_D(LuIG). Moreover, due to the 'low-T point' at T_B = 58 K [8] and the 'Schottky effect' [15], anomalies are overlapped and appear as a continuum-type over the entire C_p^{ex} curve, and the transition at T_B is then obscured. For separating both excess from each other in the C_p^{ex} curve, C_p^{sch} can be tentatively appraised experimentally using C_p(GdIG) as a baseline [14]. Herein, C_p^{RE} and $C_p^{M_{Fe}}$ in these two REIGs having similar magnetization functions are roughly comparable. In addition, as Gd^{3+} is in $_8S^{7/2}$ ground state, Schottky contribution is not expected too, and the C_p^{sch} contribution can be assumed as $C_p^{sch} \approx C_p^{total}$(TbIG) – C_p(GdIG). A small 'knee' of 17 J/K mol appears at 45 K in the C_p^{sch} curve plotted in Fig. 4 with both calculated curves of C_p^{Tb} and C_p^{sch} contributions [12]. The calculation of C_p^{sch} using the generalized equation of Moretti and Ottonello [15] was carried out. Here, the magnetic splitting energies $\bar{E}_1$ = 37.04 cm^{-1} and $\bar{E}_{1'}$ = 109.34 cm^{-1} associated with the two inequivalent sublattices C_1 and C'_1, respectively, were used [5]-[7]. Only one maximum of ~36 J/K mol is observed at 50 K in both curves with about 10 J/K mol subsisting at 150 K. C_p^{Tb} curve appears again at T > 210 K and reaches ~ 59 J/K mol at 300 K. The discrepancies between amplitudes of the maxima could be due to the inconsistency of the values of $\bar{E}_1$ and $\bar{E}_{1'}$ compared with those proposed before in low-T 36 and 50 cm^{-1}, respectively [10]. The differences above 150 K may be due to the effect of C_p^{reg} which is further fairly inaccurate in this high-T region, and the dilatation of the lattice contribution becomes not negligible [18].

4. Conclusion

In this study, magnetization, magnetic susceptibility along <111> and specific heat measurements, anomalies are reported in TbIG single crystals at temperatures in the neighbourhood (± 10 K) of the "low-T point", which are due to the simultaneous effect of the 'Schottky anomaly' near 57 K on the excess specific heat. These results imply that precise calculations of the magnetic contribution based on the NPD refined parameters of the two Tb^{3+} sublattices with a best determination of the Schottky effect based on more reliable magnetic energy levels are needed.

5. References

[1] Wang W, Zhao X, Lahoubi M and Liu J P 2015 *J. Magn. Magn. Mater.* **374** 333

[2] Lahoubi M, Wang W and Pu S 2015 *Physica* **B 476** 129

[3] Hur N *et al.* 2005 *Appl. Phys. Lett.* **87** 042901

[4] Kang T D, Standard E, Ahn K H, Sirenko A A, Carr G L, Park S, Choi Y J, Ramazanoglu M, Kiryukhin V and Cheong S-W 2010 *Phys. Rev.* B **82** 014414

[5] Lahoubi M and Ouladdiaf B 2015 *J. Magn. Magn. Mater.* **373** 108

[6] Lahoubi M 2012 Temperature evolution of the double umbrella magnetic structure in terbium iron garnet *Neutron Diffraction* ed I Khidirov (www.intechopen.com) chapter 10 pp 203–230

[7] Lahoubi M 2015 *J. Phys.: Conf. Ser.* **574** 012099

[8] Belov K P 1996 *Phys. Usp.* **39** 623

[9] Lahoubi M and Wang W 2015 *J. Magn. Magn. Mater.* **393** 437

[10] Harris A B and Meyer H 1962 *Phys. Rev.* **127** 101

[11] Tsarkhov M S, Gavrichev K S, Varazashvili V S, Gorbunov V E and Golushina L N 1990 *Russ. J. Phys. Chem.* **64** 294

[12] Parida S C, Rakshit S K and Singh Z 2008 *J. Solid State Chem.* **181** 101

[13] Varazashvili V S Tsarakhov M S and Mirianashvili T B 1997 *Russ. J. Inorg. Chem.* **42** 597

[14] Varazashvili V *et al.* 2012 Waset Conference, Paris, August 22-23

[15] Moretti R and Ottonello G1998 *Geochim. Cosmochim. Acta* **62** 1147

[16] Guillot M, Tchéou F, Marchand A, Feldmann P and Lagnier R 1981 *Z. Phys. B–Condens. Matt.* **44** 53

[17] Devyatkova E D and Tikhonov V V 1967 *Sov. Solid State* **9** 604

[18] Sayetat F 1986 *J. Magn. Magn. Mater.* **58** 334

The structural properties of $LaRO_3$ (R=Cr, Mn, Fe): a first-principles calculation

Y. R. Li[1], Z. T. Hou[1], T. X. Wang[2], Y. Li[1,*], H. Y. Liu[1], X. F. Dai[1], G. D. Liu[1]

[1] School of Material Science and Engineering, Hebei University of Technology, Tianjin 300130, China

[2] College of Physics and Information Engineering, Henan Normal University, 453007 Xinxiang, China

E-mail: liyingphy@126.com

Abstract. We investigate the structural properties of three lanthanum transition-metal oxides ($LaRO_3$, R=Cr, Mn, Fe) by the first-principles method based on density functional theory. The crystal parameters and electronic structures are calculated for three different oxides. The results show that the magnitude of lattice distortion is the strongest in $LaMnO_3$, and the slightest in $LaCrO_3$ among three compounds. The magnetic ground states of $LaCrO_3$ and $LaFeO_3$ are G-type antiferromagnet. However, for $LaMnO_3$, GGA+U fails to predict the experimental ground state, and only the internal coordinates are relaxed and the experimental lattice constants are fixed, the LDA+U can reproduce the experimental A-type antiferromagnetic state. It is found that Fe-3d electrons in $LaFeO_3$ occupy the localized orbital with the low energy, while Cr-3d and Mn-3d electrons in $LaCrO_3$ and $LaMnO_3$ occupy the relatively wide energy states due to the strong hybridization between d electrons and p electrons of oxygen atoms.

1. Introduction

Lanthanum transition-metal oxides $LaRO_3$ (R=Cr, Mn, Fe) display a complex correlation among structural, orbital, magnetic, and electronic degrees of freedom due to the existence of strongly correlated effect between 3d electrons of transition metal atoms. Over the past decades, $LaRO_3$ compounds have been widely investigated because of their technical importance and the fundamental interest in the physics of their phase transition [1, 2, 3, 4, 5, 6, 7, 8]. Among these oxides, $LaMnO_3$ had attracted considerable attention due to its colossal magnetoresistance, a large variety of phases with remarkably different structural, magnetic and transport properties [9, 10], and a large Jahn-Teller (JT) effect [11, 12]. The Jahn-Teller effect is often encountered in octahedral complexes of the transition metals, and caused mainly by the presence of molecular orbital of the 3d transition metals [13].

In the present work, we systematically investigate the structural properties and Jahn-Teller lattice distortion for three types of oxides $LaCrO_3$, $LaMnO_3$, and $LaFeO_3$ by the first-principles calculation based on density functional theory. The magnetic ground state and spin density of states of three oxides are also studied.

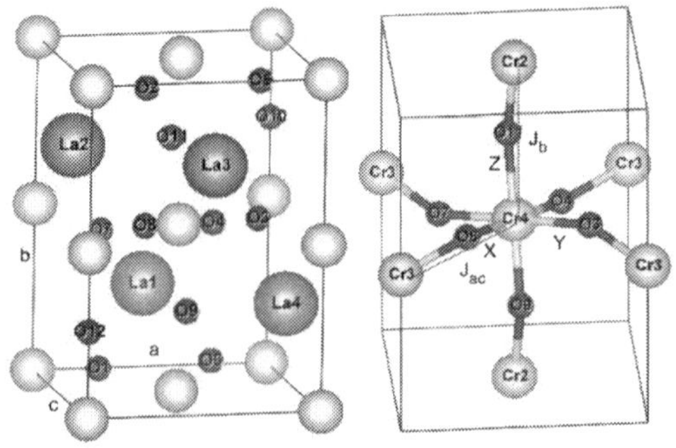

Figure 1. The orthogonal supercell model of $LaRO_3$ (left panel), where the red sphere is O atom, the green sphere is La atom, and the yellow sphere is the transition metal atom. The right panel displays a partition of orthogonal supercell (here take $LaCrO_3$ for example), where J_{ac} represents the exchange interaction parameter between the nearest-neighbor Cr3 and Cr4 atoms in the ac plane and J_b exchange interaction parameter between next nearest-neighbor Cr2 and Cr4 atoms. The X, Y, and Z represent the direction of different R-O bonds.

2. Computational details

In this paper, the first-principles calculation based on density functional theory (DFT) is performed by using the Vienna Ab initio Simulation Package (VASP). The interactions between ions and electrons are described by the projector augmented wave (PAW) method developed by Blöchl [14, 15]. The cut-off energy of the plane-wave basis is set to be 650 eV, and the Brillouin-zone integration is applied by $8 \times 8 \times 8$ Monkhorst-Pack special k-points. All atoms are fully relaxed with the total energy change less than 10^{-6} eV and the force on each ion less than 0.01 eV. The La-$5s^25p^65d^16s^2$, O-$2s^22p^4$, Cr-$3p^63d^54s$, Mn-$3p^63d^54s^2$, Fe-$3p^63d^64s^2$ are considered as valence electrons. The strong correlation effects of 3d electrons are treated by GGA+U with the Perdew-Bruke-Ernzerhof (PBE) potential. The values of U_{eff} (U-J) are taken as 5 eV for three compounds. For $LaMnO_3$, we also compute the total energy and electronic structure using the LDA exchange correlation potential.

In the calculation, we use $\sqrt{2} \times \sqrt{2} \times 2$ orthogonal supercell containing 4 transition metal atoms, 4 La atoms, and 12 oxygen atoms, as shown in Fig. 1. For the convenience of analysis, we also display a partition of orthogonal supercell in the right panel of Fig. 1, where the central atom Cr4, its neighboring atoms Cr2 and Cr3, and six oxygen atoms (O3,O4,O7,O8,O9,O11) around Cr4 are clearly seen. The experimental values of lattice constants parameters are taken as $a = 5.508$ Å, $b = 7.791$ Å, $c = 5.484$ Å for $LaCrO_3$ [16], $a = 5.705$ Å, $b = 7.703$ Å, $c = 5.536$ Å for $LaMnO_3$ [17], and $a = 5.569$ Å, $b = 7.885$ Å, $c = 5.530$ Å for $LaFeO_3$ [18].

3. Results and discussions

3.1. The magnetic ground state

To find the magnetic ground state, we first carry out the full structural optimization for $LaCrO_3$, $LaMnO_3$, and $LaFeO_3$ in the case of four magnetic configurations, i.e., ferromagnetic (FM), A-type antiferromagnetic (A-AFM), G-type antiferromagnetic (G-AFM), and C-type antiferromagnetic (C-AFM) states, as sketched in Fig.2, and calculate their total energies and magnetic moments, which are shown in Table 1. The magnetic ground states of $LaCrO_3$ and $LaFeO_3$ are found to be G-AFM state, which is in agreement with the experiment [7]. For $LaRO_3$, the result shows that $LaMnO_3$ has lower energy in FM state than in A-AFM state by GGA+U, as in previous calculations [19, 20, 21, 22]. And if only the internal coordinates with the

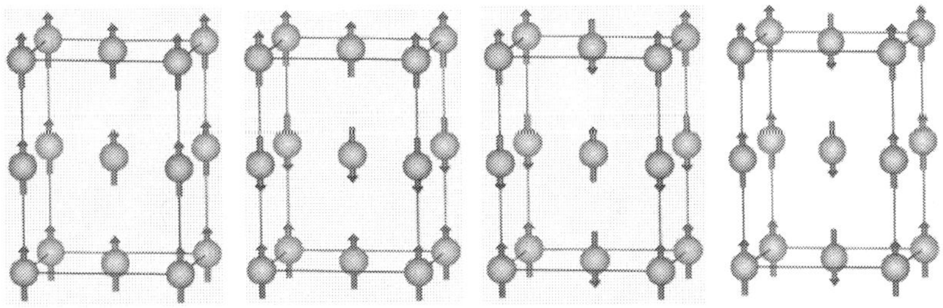

Figure 2. Four types of magnetic configurations of transition-metal atoms R in $LaRO_3$ (a) FM, (b) A-AFM, (c) G-AFM, and (d) C-AFM, where La and O atoms are removed for clarity and the green sphere represents R atoms.

experimental lattice constants are optimized by using LDA+U exchange correlation potential, we can reproduce the experimental ground state with A-type antiferromagnetic ordering. According to the energy difference, we derive the exchange coupling parameters J_{ac} (J_b) between R4 and R3 (R2)(see Fig.1 (b)), and simulate the magnetic transition temperature by Monte Carlo method [23], also given in Table 1. For comparison, the experimental transition temperature is also given in Table 1. Our simulated magnetic transition temperature $T = 79(550)$ K for $LaCrO_3$ ($LaFeO_3$) are lower than 290(750) K of the experimental values [24, 25, 26], which should be attributed to the U values and computational accuracy. For $LaMnO_3$, it is found that the magnetic moment of $3.75\mu_B$ obtained from LDA is in agreement with the experimental one of $3.7\mu_B$ [27], and the out-of-plane magnetic coupling parameter of -1.85 meV is lower than experimental one of -1.2 meV [28].

Table 1. The total energy of different magnetic states with respect to FM state, the magnetic moment of the transition-metal atom for the stable configurations, the exchange parameters, and the magnetic transition temperature of $LaCrO_3$, $LaMnO_3$, and $LaFeO_3$.

	$LaCrO_3$	$LaMnO_3$-GGA (LDA)	$LaFeO_3$
FM	0	0 (0)	0
A-AFM(meV)	-34.98	58.4 (-56.1)	-287.45
G-AFM(meV)	-113.66	305.46 (–)	-798.39
C-AFM(meV)	-80.79	256.59 (–)	-570.31
$m_R(\mu_B)$	2.95	3.9(3.75)	4.23
J_{ac}(meV)	2.214	3.86 (-1.75)	5.4
J_b(meV)	1.884	1.825 (–)	5.155
T_N(K)	79	203	550

3.2. Structure parameters

The degree of Jahn-Teller distortion can be reflected by three R-O bond lengths and R-O-R bond angles in RO_6 octahedron. In Table 2 we present the equilibrium lattice constants, normal mode [22], R-O bond lengths, O-R-O bond angle, and also R-O-R bond angles for G-AFM $LaCrO_3$, A-AFM $LaMnO_3$, and G-AFM $LaFeO_3$. The lattice constants after structure optimization are

Table 2. The calculated equilibrium lattice constants, bond lengths, and bond angles for $LaCrO_3$, $LaMnO_3$, and $LaFeO_3$.

structure	method	a(Å)	b(Å)	c(Å)	Q_2	Q_3
$LaCrO_3$	our work	5.6139	7.8942	5.5702	0.0014	0.0041
	[29]	5.523	7.801	5.527		
	[30]	5.483	7.765	5.520		
	[16]	5.501	7.791	5.484		
$LaMnO_3$	our work	5.765	7.582	5.429	0.0764	0.425
	[29]	5.623	7.814	5.528		
	[27]	5.742	7.668	5.532		
	[17]	5.7046	7.7029	5.5353		
$LaFeO_3$	our work	5.6431	7.9146	5.5772	0.0057	0.316
	[29]	5.597	7.860	5.551		
	[31]	5.565	7.854	5.557		
	[18]	5.569	7.855	5.53		
	[7]	5.563(5)	7.850(8)	5.553(3)		
	[8]	5.542	7.856	5.5596		
	$LaCrO_3$	$LaMnO_3$	$LaFeO_3$			
bond length (Å)						
R-O3(O7)	2.029	2.151	2.041			
R-O4(O8)	2.027	1.915	2.03			
R-O9(O11)	2.026	1.949	2.034			
bond angle (^{o})						
O3(O7)-R-O4(O8)	91.15	89.14	91.04			
O3(O7)-R-O11	89.49	91.82	89.35			
O4(O8)-R-O11	88.92	90.87	88.71			
R3-O7-R4	154.32	153.68	154.03			

a bit larger than the experimental data for three compounds. In $LaCrO_3$, three bond lengths, i.e., Cr4-O3 (O7), Cr4-O4 (O8), Cr4-O9 (O11), are nearly equal, and the relative ratio of bond length in the Y direction to the X (Z) direction is 0.098% (0.148%). In comparison, in $LaMnO_3$ the bond length in the Y direction is obviously larger than that in the X (Z) directions with relative ratio 10.97% (9.39%). For $LaFeO_3$ the bond length in the Y direction is slightly larger than that in the X (Z) directions with relative ratio 0.544% (0.368%).

Moreover, we measure the magnitude of JT distortion by calculating the normal mode parameters $Q_2 = (1/\sqrt{2})(X1-X4-Y2+Y5)$ and $Q_3 = (1/\sqrt{6})(2Z3-2Z6-X1+X4-Y2+Y5)$ [12, 32], where X,Y,Z directions are marked in Fig.1. The Q_2 and Q_3 of $LaMnO_3$ is larger than those of $LaCrO_3$ and $LaFeO_3$, as shown in Table 2. For three compounds, the bond angles, i.e., O3 (O7)-R-O4 (O8), O7-R-O11, O8-R-O11, have a slight deviation from the perpendicular angles each other. All above results indicate that the lattice distortion of $LaMnO_3$ is the most pronounced, $LaFeO_3$ is relatively slight, and the JT distortion of $LaCrO_3$ is almost ignorable. Additionally, the R-O-R bond angles exhibit a large deviation from the 180 degree in the perovskite structure, which also suggests the occurrence of structural distortion. The JT distortion can also be further explained by the electronic structure of 3d electrons in transition metal atoms.

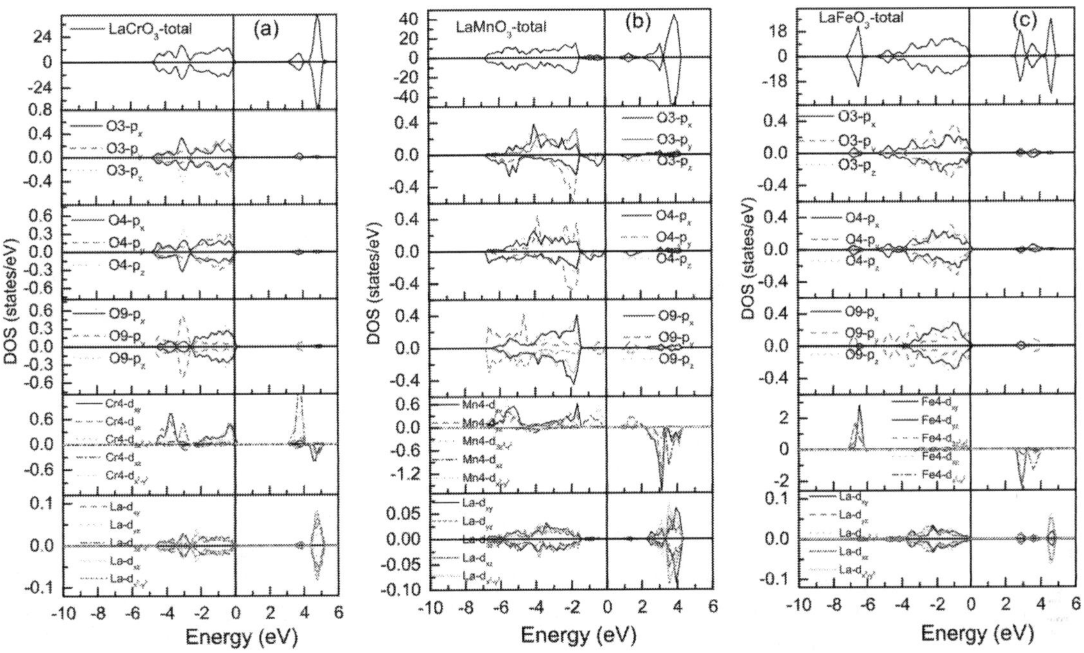

Figure 3. The total and partial density of states for O-p, Cr-d, La-d electrons in (a) $LaCrO_3$, (b) $LaMnO_3$, and (c) $LaFeO_3$.

3.3. Electronic structure

Fig.3 shows the total and partial density of states (DOS) for G-AFM-$LaCrO_3$, A-AFM-$LaMnO_3$, and G-AFM-$LaFeO_3$. It is found that three types of compounds are semiconductor. The $LaCrO_3$ has the largest energy gap of 3.1 eV, the energy gap of $LaMnO_3$ is 0.875 eV, close to other calculated results of 0.90 eV [22], and that of $LaFeO_3$ is 2.56 eV. The partial density of states of R atoms, three types of O atoms around it, and one La atom in three compounds are presented in Figs. 3 (a), 3 (b) and 3 (c), respectively, in order to clarify the orbital hybridization and ligand change around R atoms. We can see that the valence energy states are mainly contributed by the R-3d and O-2p electrons with a small contribution of La-5d electrons. There exists strong orbital hybridization between O-2p electrons and Cr (Mn)-3d electrons in $LaCrO_3$ ($LaMnO_3$).

For $LaCrO_3$ and $LaFeO_3$, three p orbitals of O3 and O4 atoms in the XY plane exhibit a slight exchange splitting, while no exchange splitting is found for three p orbitals of O9 atom along the Z direction (see Figs. 3 (a) and 3 (c)). In comparison, the p orbitals of O3 and O4 atoms in $LaMnO_3$ have obvious exchange splitting, and three p orbitals of O9 atom are also slightly asymmetry, as shown in Fig. 3 (b). Under the action of the octahedron crystal field with low symmetry, three Cr4-d electrons in $LaCrO_3$ do not apparently occupy the d_{xy},d_{yz},d_{xz} orbitals. The d electrons of Mn atom also occupy widely the five orbitals, i.e., d_{xy},d_{yz}, d_{xz}, d_{3z2-r2},d_{x2-y2}, due to the strong hybridization between 3d electrons and 2p electrons of O atoms.

In addition, it is found that the valence energy states of $LaFeO_3$ are separated into two parts. The energy states ranging from -8 eV to -6 eV are contributed by Fe-3d electrons and small O-2p electrons, and those from -5 eV to 0 eV are contributed by O-2p electrons and La-5d electrons. Compared with that of $LaCrO_3$ and $LaMnO_3$, the orbital hybridization of $LaFeO_3$ between O-2p and Fe-3d electrons becomes weak. The Fe-d electrons orbital in $LaFeO_3$ is very localized and has a small energy scale ranging from -7.1 eV to -6.1 eV due to small orbital hybridization between Fe-3d electrons and O-2p electrons.

4. Conclusions

In summary, we have investigated the magnetic ground state, structure parameters, electronic density of states in $LaCrO_3$, $LaMnO_3$ and $LaFeO_3$ by the first-principles calculation. The results show that the magnetic ground states of $LaCrO_3$ and $LaFeO_3$ are G-type antiferromagnetic state, which is in agreement with the experimental results. For $LaMnO_3$, it is found that the ferromagnetic ground state has lower energy than the antiferromagnetic state in the case of optimizing both internal coordinates and lattice constants based on GGA+U, and the A-type antiferromagntetic state can be obtained if only optimizing the internal coordinates with the experimental lattice constants based on LDA+U. The lattice distortion of $LaMnO_3$ is the most pronounced among three compounds. Compared with the d electrons in $LaCrO_3$ and $LaMnO_3$, the d electrons in $LaFeO_3$ have small energy splitting and occupy the very localized orbital.

5. Acknowledgement

This work is funded by the National Nature Science Foundation of China (Grant Nos. 11204064 and U1304518)

References

[1] Jiang S, Liu L, Ong K, Wu P, Li J, Pu J 2008 *Power Sources* **176** 82
[2] Nithya V, Immanuel R, Senthilkumar S, Sanjeeviraja C, Perelshtein I, Zitoun D, Selvan R 2012 *Mater. Res. Bull.* **47** 1861
[3] Nakayama S 2001 *Mater. Sci.* **36** 5643
[4] Zhou J, Alonso J, Muoz A, Fernndez-Daz M, Goodenough J 2011 *Phys. Rev. Lett.* **106** 057201
[5] Scafetta M, Cordi A, Rondinelli J, May S 2014 *Phys.Condens. Matter* **26** 505502
[6] Taran O, Ayusheev A, Ogorodnikova O, Prosvirin I, Isupova L, Parmon V 2016 *Applied Catalysis B: Environmental* **180** 86
[7] Mahapatra A, Mitra A, Mallick A, Ghosh M, Chakrabarti P 2016 *Materials Letters* **169** 160
[8] Chanda S, Saha S, Dutt A, Irfan B, Chatterjee R, Sinha T.P. 2015 *Journal of Alloys and Compounds* **649** 1260
[9] Zenia H, Gehring G,Temmerman W 2005 *New J. Phys.* **7** 257
[10] Banach G, Temmerman W 2004 *Phys. Rev. B* **69** 054427
[11] Dagotto E, Hotta T, Moreo A 2001 *Phys. Rep.* **344** 1
[12] Hou Y, Xiang H, Gong X 2014 *Phys.Rev.B* **89** 064415
[13] Brien O, Mary C, Chancey C 1993 *Am J.Phys.* **61** 688
[14] Blochl P 1994 *Phys.Rev.B* **50** 17953
[15] Kresse G, Joubert J 1999 *Phys.Rev.B* **59** 1758
[16] Tolochko S, Kononyuk I, Zonov Y, Ivashkevich L 1987 *Inorg. Mater.* **23** 743
[17] Krogh I, Krogh A, Norby P, Skou E 1994 *Solid State Chem.* **113** 320
[18] Iguchi E, Jung W 1994 *Phys.Soc.Jpn.* **63** 3078
[19] Sawada H, Morikawa Y, Terakura K, Hamada N 1997 *Phys.Rev.B* **56** 12154
[20] Sawada H, Terakura K 1998 *Phys.Rev.B* **58** 6831
[21] Kotomin E, Evarestov R, Mastrikov Y, Maier J 2005 *Phys.Chem* **7** 2346
[22] Hashimoto T, Ishibashi S, Terakura K 2010 *Phys.Rev.B* **82** 045124
[23] Li Y, Liu B 2006 *Phys.Rev.B* **73** 174418
[24] Tiwari B, Dixit A, Naik R, Lawes G and Ramachandra Rao M 2014 *Materials Research Express* **2** 2
[25] Hearne G, Pasternak M 1995 *Phys.Rev.B* **51** 11495
[26] Mahapatra A, Mitra A, Mallick A, Ghosh M, Chakrabarti P 2016 *Materials Letters* **169** 160
[27] Elemans J, Van Laar B, Van Der Veen K,Loopstra B 1971 *J.Solid State Chem* **3** 238
[28] Moussa F, Hennion M, Rodriguez-Carvajal J, Moudden H, Pinsard L, Revcolevschi A 1996 *PRB* **54** 15149
[29] Wang Y and Cheng H. 2013 *J. Phys. Chem. C* **117** 2106
[30] Khattak C, Cox D 1977 *Mater. Res. Bull.* **12** 463
[31] Dann S, Currie D, Weller M, Thomas M, Al-Rawwas A 1994 *J.Solid State Chem* **109** 134
[32] Kanamori J 1960 *J.Appl.Phys.* **31** 14S

IOP Conf. Series: Journal of Physics: Conf. Series **827** (2017) 012016 doi:10.1088/1742-6596/827/1/012016

Magnetism induced by Ga vacancy in rare earth R(R=Gd,Eu,Tm)doped GaN

Z.T. Hou[1], Y.R. Li[1], T.X. Wang[2], Y. Li[1,*], H.Y. Liu[1], X.F. Dai[1], G.D. Liu[1]

[1] School of Material Science and Engineering, Hebei University of Technology, Tianjin 300130, China

[2] College of Physics and Information Engineering, Henan Normal University, 453007 Xinxiang, China

E-mail: `liyingphy@126.com`

Abstract. In recent years, GaN doped with rare earth has attracted much attention due to its potential application in spintronic devices and optoelectronic devices. Based on the density functional theory, we investigate the magnetic moment, formation energy, and electronic structure in R (R= Gd, Eu, Tm) doped GaN semiconductors. We focus on the contribution of Ga vacancy to the magnetism, and calculate the formation energy of different Ga vacancies in the presence of Gd, Eu, or Tm dopants, and that of Gd, Eu, Tm dopants in native defects of Ga vacancy. The possible stable defect structures in GaN are given according to their formation energy. It is found that the Ga vacancies prefer to form cluster, and the formation energy of concentrated Ga vacancies is low compared to separated Ga vacancies. The 5d electrons of rare earth as well as 4f electrons have a larger contribution to the magnetism of GaN:R with Ga vacancy than without Ga vacancy. In addition, intermediate bands were observed in GaN:R with intrinsic defects, which possibly opens the potential application of R-doped semiconductors in the third generation high efficiency photovoltaic devices.

1. Introduction

Over the past decades, rare-earth doped semiconductors have received wide attention because of their possible applications in spintronic and optoelectronic devices. In 1989, Favennec et al.[1] demonstrated that the thermal queching of the luminescence in the erbium doped semiconductors decreases with the increase of band gap of the host materials. As one of typical wide band gap semiconductors, GaN is an ideal host candidate for rare-earth ions because of its high field transport characteristics and incorporating with the rare-earth ions very well. Recently, Gd-doped GaN (GaN:Gd) has attracted considerable interest due to the characters of its high temperature ferromagnetism and colossal magnetic moment[2, 3, 4, 5]. However, the microscopic mechanism of ferromagnetism in GaN:Gd still remains controversial since the ferromagnetism was discovered first in 2000[6]. Dhar et al.[7, 8, 9] pointed out that the ferromagnetic coupling in the GaN:Gd system cannot be explained simply in terms of direct, double, or superexchange between Gd atoms since the average Gd-Gd distance is too large. Gohda et al.[10] proposed that the ferromagnetism of GaN:Gd should be derived from the occurrence of cation vacancy, and Thiess et al. found that Ga vacancy cluster in GaN:Gd has the stable structure and may induce the enlargement of magnetic moment[11].

IOP Conf. Series: Journal of Physics: Conf. Series **827** (2017) 012016 doi:10.1088/1742-6596/827/1/012016

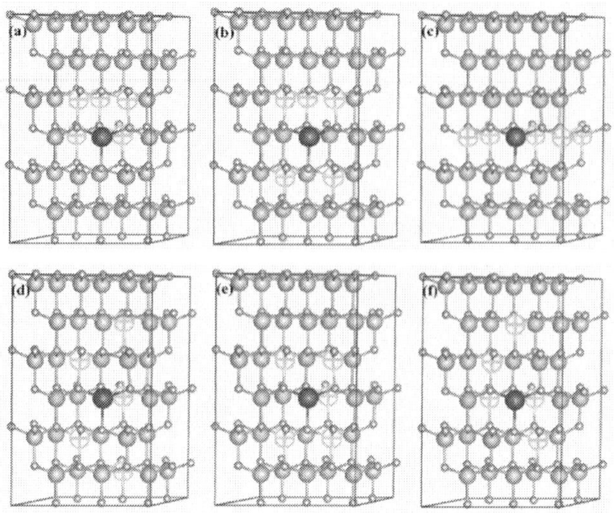

Figure 1. Six different defect complex configurations of GaN : R + V_{Ga} constituted by one R dopant and five Ga vacancies. Ga, N, RE ions and Ga vacancies are represented by green, gray, red and empty yellow balls, repectively.

In this paper, we investigate the magnetic moment, formation energy, and electronic structures of R (R=Gd, Eu, Tm) dopants and their complexes with cation Ga vacancies in wurtzite GaN by the density functional theory. We first calculate the crystal parameters of isolated Gd dopant and defect complexes with one Ga or N vacancy. Then five Ga vacancies are systematically considered in GaN:R to explore the origin of magnetism, and the formation energies of different defect complex configurations are discussed. Finally, we analyze the electronic structures of the most stable defect configurations of Gd, Eu, Tm-doped GaN.

2. Computational details

In this work, all calculations are performed with the Vienna ab-initio simulation package (VASP) by using the pseudopotential plane-wave scheme[12]. The interactions between ions and electrons are described by the projector augmented wave (PAW) method developed by Blöchl [13, 14]. The cutoff energy of the plane wave basis set is 500 eV, and the Brillouin-zone integration is performed by 5×5×3 Mokhorst-Pack special k-points. All atoms are fully relaxed with the total energy change less than 10^{-5} eV and the force on each ion less than 0.03 eV/Å. The Ga $4s^2 4p^1$, N $2s^2 2p^3$, Gd $5s^2 5p^6 4f^7 5d^1 6s^2$, Eu $5s^2 5p^6 4f^6 5d^1 6s^2$, and Tm $5s^2 5p^6 4f^{12} 5d^1 6s^2$ are considered as valence electrons. The strong correlation effect of the R-$4f$ electrons is treated by the local spin density approximation with Hubbard-U corrections (LSDA+U) with $U = 9.2, 8.506, 10.082$ eV and $J = 1.2, 1.109, 1.317$ eV for Gd, Eu, Tm taken from Ref.[15].

To explore the defect properties, a $3 \times 3 \times 2$ GaN supercell containing 72 atoms ($Ga_{36}N_{36}$) is used. We substitute one R for one Ga atom, corresponding to a 2.78% R doping concentration. Here, one Ga-site is considered for R substitution because the R_{Ga} has been both experimentally and theoretically confirmed to be the most stable configuration in all states[16, 17, 18]. In this paper we focus on six different defect complexes constituted by replacing one Ga atom with one R atom and simultaneously removing five Ga atoms on different neighboring positions, i.e., GaN:R+5V_{Ga}, as illustrated in Fig. 1.

Table 1. Relaxed crystal parameters and calculated magnetic moments of three defect systems, i.e., GaN, GaN:Gd, and GaN:Gd+V_{Ga}.

	a(Å)	c(Å)	c/a	$m_{Gd}(\mu_B)$	$m_{tot}(\mu_B)$
GaN	3.211	5.238	1.631	-	-
GaN:Gd	3.282	5.737	1.748	7.173	7.045
GaN:Gd+V_{Ga}	3.292	5.383	1.635	7.171	9.110

3. Results and discussions

In table 1 we show the relaxed crystal parameters and the calculated magnetic moments for two simple defect systems, i.e., the isolated dopant Gd (GaN:Gd) and the defect complexes (GaN:Gd+V_{Ga}) constituted by one Gd and one Ga vacancy, and also pure GaN for comparison. For GaN:Gd, the lattice constants a and c expand about 2.2% and 9.5% compared with the pure GaN due to the ionic radius of Gd (0.94 Å) larger than that of Ga (0.62 Å). For the defect complexes, the lattice constant c shrinks compared to isolated Gd dopant due to the presence of Ga vacancy, which agrees with the previous study[19]. The magnetic moment of Gd atom is about 7.1 μ_B, and Ga vacancy can induce $2\mu_B$. Therefore, the Ga vacancies are possibly responsible for the enlarged magnetic moment of GaN:Gd. In the following, the various defect complexes containing dopants R and five Ga vacancies are theoretically investigated for exploring the effect of vacancy on the magnetism.

Table 2. Magnetic moment(m_R) of R atom and the total magnetic moment(m_{tot}) in GaN:Gd+5V_{Ga}, GaN:Eu+5V_{Ga}, and GaN:Tm+5V_{Ga}. The unit of magnetic moment is μ_B.

	GaN:Gd+5V_{Ga}		GaN:Eu+5V_{Ga}		GaN:Tm+5V_{Ga}	
	m_{Gd}	m_{tot}	m_{Eu}	m_{tot}	m_{Tm}	m_{tot}
a	8.096	17.178	6.444	16.756	2.226	11.311
b	8.258	17.428	6.392	16.829	2.205	12.735
c	11.153	21.110	1.652	10.872	0.218	9.127
d	7.815	18.744	6.465	17.089	2.211	12.986
e	8.111	18.742	-0.174	10.655	1.924	12.690
f	8.209	18.899	4.330	14.885	0.222	10.859

Table 2 lists the magnetic moments of six defect complex configurations for GaN: R+5V_{Ga} (see Fig. 1). It can be seen that the largest magnetic moment is obtained for Gd in configuration-c, Eu in configuration-d, and Tm in configuration-a. For the total magnetic moment, the largest value of GaN:Gd+5V_{Ga} occurs in configuration-c, where all vacancies lie in the same cation plane as Gd, while GaN:Eu+5V_{Ga} and GaN:Tm+5V_{Ga} have the largest total magnetic moments in configuration-d. In addition, compared with Gd atom, the magnetic moments of Eu (Tm) atoms rather strongly depend on the Ga vacancy configurations. This should be attributed to the different crystal field effect on the delocalized f states of Eu (Tm) atoms for different Ga vacancy configurations. We further calculate the formation energies of six defect complex configurations to explore their energetic stability. $\Delta E_R(V)$ represents the formation energy of

IOP Conf. Series: Journal of Physics: Conf. Series **827** (2017) 012016 doi:10.1088/1742-6596/827/1/012016

Ga vacancies in the presence of dopants and $\Delta E_V(R)$ is that of dopant atom in the native Ga vacancies. They are respectively expressed as[20]

$$\Delta E_V(R) = E_{R+V} - E_V + \mu_{\rm Ga} - \mu_{\rm R}, \tag{1}$$

and

$$\Delta E_R(V) = E_{R+V} - E_R + n_{\rm Ga}\mu_{\rm Ga}, \tag{2}$$

where E_{R+V} is the total energy of the supercell containing R dopants and Ga vacancies. E_R is the total energy of GaN:R without Ga vacancies, and E_V is the total energy of GaN with Ga vacancies and without R dopants. μ_i is the corresponding chemical potential of atom i (i=Ga, R) referenced to elemental solid with energy $E(i)$, and n_{Ga} ($n_{Ga} > 0$) indicates the number of Ga atoms removed from the supercell.

Table 3. Formation energies of Gd, Eu, Tm dopants in the native Ga vacancy (Eq. 1) and that of Ga vacancies in the presence of R dopants (Eq. 2).The unit of formation energy is eV

	GaN:Gd+5$V_{\rm Ga}$		GaN:Eu+5$V_{\rm Ga}$		GaN:Tm+5$V_{\rm Ga}$	
	$\Delta E_{\rm V}$(Gd)	$\Delta E_{\rm Gd}$(V)	$\Delta E_{\rm V}$(Eu)	$\Delta E_{\rm Eu}$(V)	$\Delta E_{\rm V}$(Tm)	$\Delta E_{\rm Tm}$(V)
a	-0.081	38.874	-1.528	35.439	-1.584	36.460
b	0.287	39.312	-1.571	35.695	2.579	40.851
c	-16.545	22.448	4.673	41.869	-0.497	37.718
d	1.163	40.448	0.154	37.979	-0.368	38.451
e	-0.310	39.019	4.382	41.878	-0.157	38.397
f	-0.043	39.493	1.403	39.171	-1.208	37.591

In table 3 we list the above two types of formation energies. It is found that the Ga vacancies in GaN doped with R are very difficult to form, and the R doping in GaN with the native Ga vacancies is relatively easy. According to the formation energy $\Delta E_{\rm V}$(R), GaN:Gd+5$V_{\rm Ga}$-c has the lowest formation energy, and GaN:Eu+5$V_{\rm Ga}$-b and GaN:Tm+5$V_{\rm Ga}$-a have the lowest formation energy. For the three configurations, Ga vacancies have a dense distribution, which is in qualitative agreement with the other work[11]. Thus our calculated results show that GaN:Gd+5$V_{\rm Ga}$-c, GaN:Eu+5$V_{\rm Ga}$-b, GaN:Tm+5$V_{\rm Ga}$-a are the stablest defect structures among all defect structures studied.

Fig. 2 shows the total and partial density of states (PDOS) of GaN:Gd+5$V_{\rm Ga}$-c, GaN:Eu+5$V_{\rm Ga}$-b, GaN:Tm+5$V_{\rm Ga}$-a, and also GaN:Gd for comparison. We see that the conduction band of GaN doped with R is primarily attributed to N-2p, Ga-4s, Ga-4p, and R-4f states. The magnetic moment of R atoms is mainly contributed by their d and f states. The N atoms around Ga vacancies can induce the localized levels close to the Fermi level in the down spin channel, which leads to the magnetic moment, as shown in Figs. 2(b)-2(d). In contrast to that of GaN:Gd+5$V_{\rm Ga}$, the f states of GaN:Eu+5$V_{\rm Ga}$ and GaN:Tm+5$V_{\rm Ga}$ are relatively delocalized and are closer to Fermi level, so that the crystal field effect of f states is more significant for GaN:Eu+5$V_{\rm Ga}$ and GaN:Tm+5$V_{\rm Ga}$ than GaN:Gd+5$V_{\rm Ga}$.

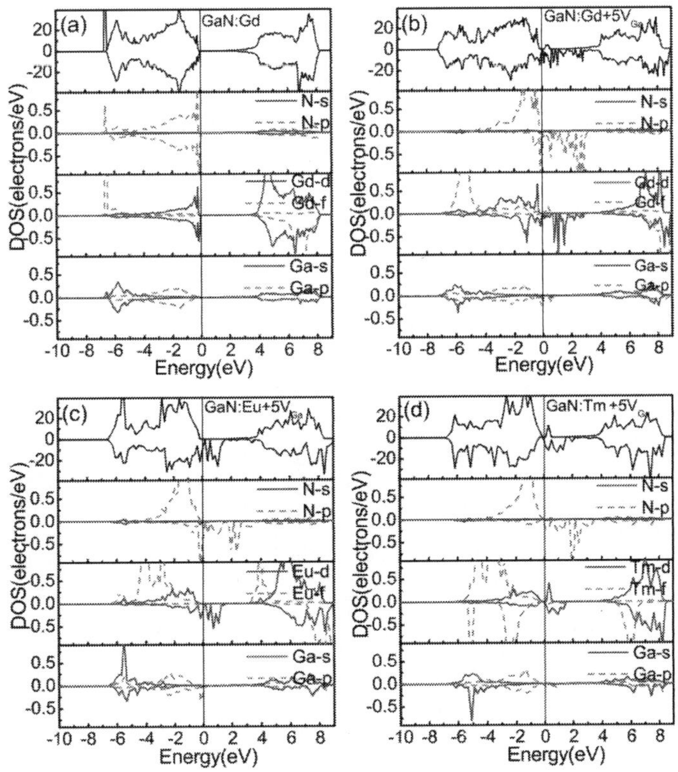

Figure 2. The density of states (DOS) of (a) isolated substitutional GaN:Gd, and of defect complexes (b) GaN:Gd+5V_{Ga}, (c) GaN:Eu+5V_{Ga}, and (d) GaN:Tm+5V_{Ga} with the stable defect structures.

4. Conclusions

In summary, we have performed an investigation of the magnetic moment, formation energy, and electronic structure of GaN:R (R=Gd, Eu, Tm) with Ga vacancy defects by using DFT calculation. Our calculated results show that the introduction of R dopants in GaN with the native Ga vacancies is relatively easy, but the formation of Ga vacancies in the presence of R dopants is very difficult. The magnetic moment of rare earth R in GaN:R is strongly dependent on the defect complex configurations. The f states in GaN:Eu+5V_{Ga} and GaN:Tm+5V_{Ga} are more delocalized and closer to the Fermi level than that in GaN:Gd+5V_{Ga}.

5. Acknowledge

This work is funded by the National Nature Science Foundation of China (Grant Nos. 11204064 and U1304518)

References

[1] Favennec P, Haridon H, Salvi M, Moutonnet D and Guillou Y 1989 *Electron. Lett.* **25** 718
[2] Steckl A, Heikenfeld J, Lee D, Garter M, Baker C, Wang Y and Jones R 2002 *IEEE J.Select.Top.Quantum Electron* **8** 749
[3] Jones R 2006 *Opt.Mater.* **28** 718

[4] Chen S, Dierre B, Lee W, Sekiguchi T, Tomita S, Kudo H and Akimoto K 2010 *Appl.Phy.Lett.***96** 181901.
[5] Jiang L, Wang X, Xiao H, Wang Z, Yang C and Zhang M 2011 *Appl.Phys.A* **104** 429
[6] Teraguchi N, Suzuki A, Nanishi Y, Zhou Y K, Hashimoto M, Asahi H 2000 Solid State Commun 122 651
[7] Dhar S, Prez L, Brandt O, Trampert A, Ploog K, Keller J and Beschoten B 2005 *Phys.Rev.B* **24** 245203
[8] Dhar S, Brandt O, Ramsteiner M, Sapega V and Ploog K 2005 *Phys.Rev.Lett.* **03** 7205
[9] Dhar S, Kammermeier T, Ney A, Perez L, Ploog K, Melnikov A and Wieck A 2006 *Appl.Phys.Lett* **06** 2503
[10] Gohda Y, Atsushi Oshiyama 2008 Phys.Rev. B 78 161201
[11] Thiess A, Dederichs P H, Zeller R, Blugel S, Lambrecht W R L 2012 Phys.Rev.B 86 180401
[12] Kresse G andFurthmuller J 1996 *Phys.Rev.B* **54** 11169
[13] Blochl P 1994 *Phys.Rev.B* **50** 17953.
[14] Kresse G and Joubert D 1999 *Phys.Rev.B* **59** 1758
[15] Larson P, Walter R. L. Lambrecht, Athanasios Chantis and Mark van Schilfgaarde 2007 *Phys.Rev.B* **75** 045114
[16] Sanna S, Schmid W, Frauenheim T and Gerstmann U 2009 *Phys.Rev.B* **80** 104120
[17] Lozykowski H 1993 *Phys.Rev.B* **48** 17758
[18] Filhol J, Jones R, Shaw M and Briddon P 2004 *Appl.Phys.Lett.* **84** 2841
[19] Thiess A, Blugel S, Dederichs P, Zeller R and Lambrecht W 2015 *Phys.Rev.B* **92** 104418
[20] Hou Z,Wang X, Ikeda T,Terakura K,Oshima M,Kakimoto M and Miyata S 2012 *Phys.Rev.B* **85** 165439

Effect of magnetic field strength on M_6C carbide precipitation behavior in $W_6Mo_5Cr_4V_3$ high speed steel during tempering

Y Wu[1], Z W Zhang [2], H H Li[2] and X Zhao*[2]
[1]Research Academy, Northeastern University, Shenyang 110819, Liaoning, China
[2] Key Laboratory for Anisotropy and Texture of Materials (Ministry of Education), Northeastern University, Shenyang 110819, Liaoning, China
*Corresponding author: zhaox@mail.neu.edu.cn

Abstract. Effect of high magnetic field strength on M_6C carbide precipitation morphology in $W_6Mo_5Cr_4V_3$ high speed steel was investigated. Results showed that at low and medium tempering temperatures, the high magnetic field significantly affects the precipitation morphology of M_6C carbides and shows strong spheroidization. This effect increases with the enhancement of the magnetic field strength. At high tempering temperature, the high magnetic field has no obvious effect on M_6C carbide precipitation behavior.

1. Introduction

In recent years, the influence of high magnetic field on phase transformations in steels has attracted much research interest, such as the study of magnetic field on martensite transformation [1-3], ferrite transformation [4-6], and pearlite transformations [7-9].

Carbides play an important role in the strengthening and toughening of microalloyed steels. Previous research has shown that the high magnetic field (10T or 12T or 14T) can affect the carbide growth behavior [10, 11] and change the carbides precipitation sequence [12-15]. These researches make great progress on the understanding of high magnetic field mechanism on carbide precipitation in steels. However, rare work has been involved in studying the effect of magnetic field strength on carbide precipitation behavior. In order to further investigate the mechanism of high magnetic field on carbide precipitation behavior during tempering, in this work, $W_6Mo_5Cr_4V_3$ high speed steel was selected and treated with different magnetic field strength. The effect of magnetic field strength on precipitation behavior of M_6C carbide was mainly focused.

2. Experimental

The alloy used in this study is an annealed HIP powder metallurgy $W_6Mo_5Cr_4V_3$ high speed steel. The chemical composition is (wt %): 1.24 C, 3.99 Cr, 5.25 Mo, 6.13 W, 3.01 V, 0.55 Si, 0.28 Mn, 0.18 Ni, 0.02S, 0.02 P, 0.14 Cu, 79.19 Fe. Specimens were first austenitized at 1180 °C for 5 min in a KSX-6-14 box-type furnace, and then immediately quenched in quenching oil at 20°C. The quenched specimens were then triple tempered at 200°C, 400°C and 560°C for 80min respectively, with different magnetic field strength (0T, 1T, 6T, 12T). In the case of the magnetic field tempering, the specimens were placed in the central (zero magnetic force) region of the furnace with their longitudinal direction parallel to the magnetic field direction (MD). The high magnetic field was applied during the whole heating, isothermal holding processes. The microstructure and composition of the carbides is analyzed by scanning electron microscopy and Electron Probe Microanalysis (EPMA), respectively. In order to study the influence of high magnetic field on the precipitation morphology of M_6C carbides, 15 microstructure images (×2000) of different areas are randomly selected. The Image-pro Plus and Adobe Photoshop CS software are used to analyze statistically the spherical carbides (axis ratio$<$1.2) proportion of M_6C carbides.

3. Results and discussion

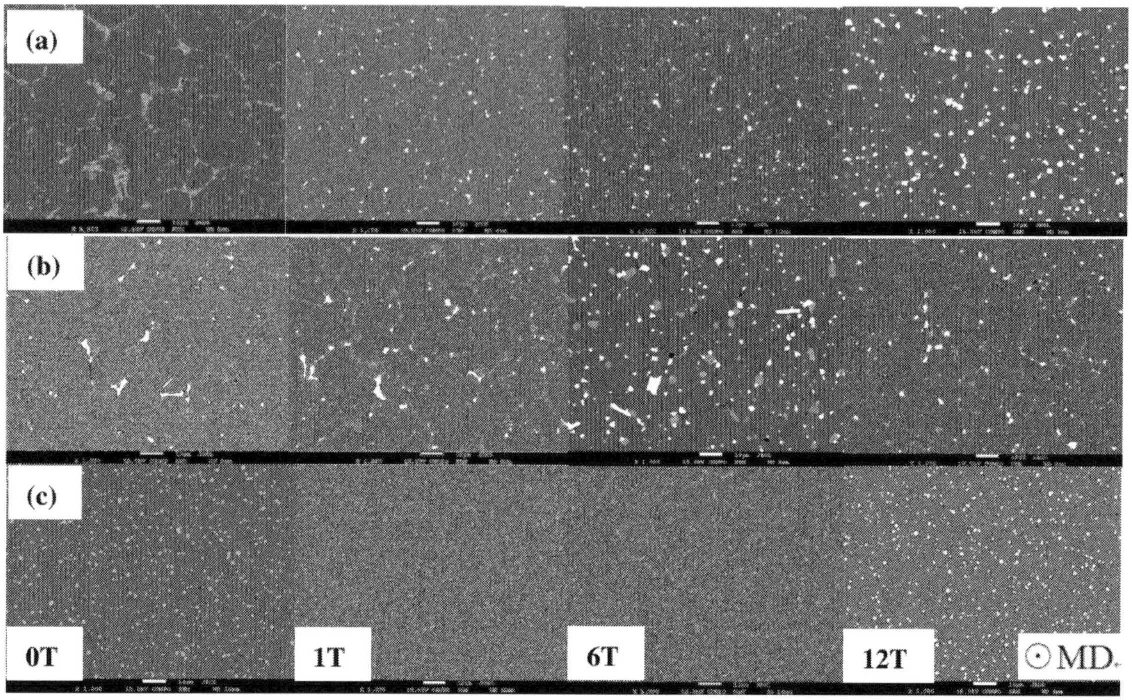

Figure 1. Microstructure of specimens tempered at 200°C (a), 400°C (b) and 560°C (c) with different magnetic field strength.

Figure 1 shows the microstructures of $W_6Mo_5Cr_4V_3$ high speed steel tempered at 200°C, 400°C and 560°C with different magnetic field strength, respectively. In the figures, the magnetic field direction is perpendicular to the paper. On the basis of EPMA analysis (table 1), both the white angular-like and the ball-block-like carbides in figures are M_6C carbides. Here, M represents the metallic elements of Fe, Mo and W. From Figure 1(a) we can see that when tempered at 200°C with magnetic field of 0T, M_6C carbides precipitate as coarse angular-like along grain boundaries and ball-block-like in intracrystalline. When tempered with a magnetic field of 1T, 6T and 12T, the coarse angular-like M_6C carbides precipitating along grain boundaries disappear, and all the M_6C type carbides precipitate as ball-block-like and distribute dispersedly. From Figure 1(b) we can see that when tempered at 400°C with magnetic field of 0T, M_6C type carbides also precipitate as coarse angular-like along grain boundaries and ball-block-like in intracrystalline. With the increasing of magnetic field strength, the coarse angular-like M_6C carbides precipitating along grain boundaries decrease gradually; when tempered with a magnetic field of 12T, the coarse angular-like M_6C carbides disappear, and the M_6C type carbides precipitate as ball-block-like and distribute dispersedly. From Figure 1(c) we can see that when tempered at 560°C, the precipitation morphology of M_6C carbides is basically unchanged with the increasing of magnetic field strength, and all the M_6C carbides precipitate as ball-like. The above results indicate that at low and medium tempering temperatures, the high magnetic field significantly affects the precipitation morphology of M_6C carbides and leads to strong spheroidization, the stronger the magnetic field strength, the stronger the spheroidization effect. At high tempering temperature, the high magnetic field has no obvious effect on M_6C carbide precipitation morphology. The effect of high magnetic field on the precipitation morphology of M_6C carbides becomes weak with the increasing of the tempering temperature.

There is no obvious effect on the precipitation sites of M_6C carbides, and the M_6C carbides can precipitate at original austenite boundaries and in grain interior under different magnetic field strength.

Element	Atomic%
C	13.44
V	4.95
Cr	4.61
Fe	44.46
Mo	17.01
W	15.53

Table 1. The EPMA analysis of M_6C carbides in tempered specimens.

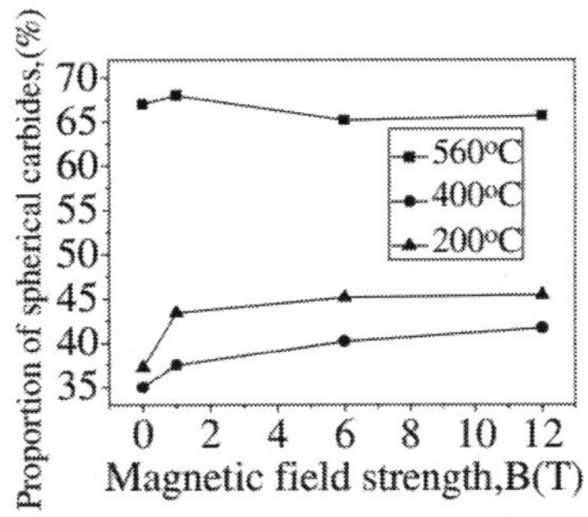

Figure 2. The proportion of spherical carbides in tempered specimens.

Figure 2 shows the spherical carbide proportion of M_6C carbides in specimens tempered with different magnetic field strength. From Figure 2 we can see that, the proportion of spherical M_6C type carbides increases remarkably with the increasing of magnetic field strength when tempered at 200°C and 400°C. When tempered at 560°C, the proportion of spherical M_6C type carbides is basically unchanged with the increasing of magnetic field strength.

Without magnetic fields, M_6C on grain boundaries shows a strong tendency to grow along boundaries, and coarse angular-like M_6C are eventually formed. This growth is energetically favorable as it can make use of the existing interfacial energy to lower the additional energy required for the formation of new carbide/matrix interfaces. But this growth tendency is not favored by the magnetic field. When M_6C carbides grow, the interfacial area becomes larger. If there is no magnetic field, the interfacial energy of M_6C /matrix is only determined by the crystal structure and composition differences between the two phases; conversely, if there is a magnetic field, the interfacial energy resulting from the difference in magnetization between the two phases should also be taken into account. For a solid phase, when magnetized by the applied magnetic field, the Gibbs free energy can be decreased. Therefore, the Gibbs free energy values for M_6C and matrix on both sides of the M_6C /matrix interface can be reduced by an amount corresponding to their magnetization degree. Conversely, the energy level of the interface remains unchanged, as the magnetization of interface is very low due to the disorderly atom arrangement and the large number of crystal defects. Thus, the interface energy increases relative to the grain interior, as suggested in Ref. [10]. Since the magnetic field can obviously raise the M_6C/matrix interfacial energy, the shape of M_6C that has minimum interface area is advantageous to minimize the final total interfacial energy. Therefore, the sphere-or particle-like M_6C is most favorable. In addition, the magnetostriction of M_6C

is different from that of the matrix. The growth of M_6C will result in a difference in volume and therefore an increase in strain energy. According to the phase transformation theory, the strain energy resulting from this difference in volume depends on the shape of precipitates. When the hardness of the precipitates is much higher than that of the matrix, spherical precipitates generate the lowest strain energy [16]. Under the two effects of the magnetic field, the shape of particle-like M_6C that has the minimum total interfacial area and minimum magnetostrictive strain energy is most favorable. Therefore, the magnetic field promotes the spheroidization of M_6C type carbides, and this effect increases with the enhancement of the magnetic field strength. With the increasing of tempering temperature, the influence of temperature on Gibbs free energy increases, resulting in the weakening of the influence of magnetic field. So at high tempering temperature (560°C), the high magnetic field almost has no obvious effect on M_6C carbide precipitation morphology.

4. Conclusions

Effect of high magnetic field strength on M_6C carbide precipitation behavior in $W_6Mo_5Cr_4V_3$ high speed steel has been investigated, which leads to the following results: (1) At low and medium tempering temperatures, the high magnetic field obviously affects the precipitation morphology of M_6C carbides and shows strong spheroidization, the stronger the magnetic field strength, the stronger the spheroidization effect. (2) The effect of high magnetic field on the precipitation morphology of M_6C carbides becomes weak with the increasing of the tempering temperature.

5. Acknowledgments

The authors are grateful to the financial support for this work from the National Natural Science Foundations of China (Grant No. 51271052) and the “111” Project (Grant No. B07015).

References

[1] Shimozono T, Kohno Y, Konishi H, et al 1999 *Mater. Sci. Eng. A* **A273-275** 337
[2] San Martín D, D Aarts K W, et al 2008 *J. Magn. Magn. Mater.* **320** 1722
[3] San Martín D, Van Dijk N H, Jiménez-Melero E, et al 2010 *Mater. Sci. Eng. A* **527** 5241
[4] Zhang Y D, He C S, Zhao X, Esling C and Zuo L 2004 *Adv. Eng. Mater.* **6(5)** 310
[5] Shimotomai M, Maruta K, Mine K and Matsui M 2003 *Acta Mater.* **51** 2921
[6] Zhang G H, Enomoto M., Hosokawa N, et al 2009 *J. Magn. Magn. Mater.* **321** 4010
[7] Wu C Y, Li T J, Wen B and Jin J Z 2004 *J. Mater. Sci. Technol.* **39** 1129
[8] Zhang G H and Enomoto M 2010 *J. Mater. Sci. Technol.* **26** 269
[9] Zhang X X, Zhang Y D, Gong M L, et al 2012 *J. Magn. Magn. Mater.* **324** 4188
[10] Zhang Y D, Gey N, He C S, Zhao X, Zuo L and Esling C 2004 *Acta Mater.* **52** 3467
[11] Xia Z X, Zhang C, Lan H, Liu Z Q and Yang Z G 2011 *Mater. Lett.* **65** 937
[12] Zhou Z N, Wu K M 2009 *Scripta Mater.* **61(7)** 670
[13] Hou T P and Wu K M 2012 *Scripta Mater.* **67** 609
[14] Hou T P and Wu K M 2013 *Acta Mater.* **61** 2016
[15] Hou T P, Li Y, Zhang Y D and Wu K M 2014 *Metall. Mater. Trans. A* **45A** 2553(1-9)
[16] Hsu T Y *Theory of phase transformation*, Science Press of China, Beijing, 1988

IOP Conf. Series: Journal of Physics: Conf. Series **827** (2017) 012018 doi:10.1088/1742-6596/827/1/012018

Magnetic Properties and Exchange Bias Effect of Cu and Cr doped Sm_2Co_{17} Alloy Film

Dashun Dong, Qingqing Fang*, Wenwen Wang, Jinjin Yang

School of Phys. & Mater. Sci., Anhui University, Hefei, 230039, China

*Corresponding author: physfangqq@126.com

Abstract Cu and Cr-doped Sm_2Co_{17} films were synthesized using pulsed laser deposition method with the substrate of Si(100), and our results show that Cu and Cr incorporation can reduce the grain size and increase the coercivity of Cu and Cr-doped Sm_2Co_{17} films. Furthermore, incorporations of Cu and Cr will introduce a positive effect for the exchange bias of Cu and Cr-doped Sm_2Co_{17} films. The magnetic measurement of external magnetic field and zero-field-cooled (FC-ZFC) under the temperature ranging from 10K to 300 K prove that the critical value of FM-AFM transform is ~80 K in the Cu and Cr-doped Sm_2Co_{17} films, which showed that the exchange bias origins from the coupling competition between ferromagnetic phase and antiferromagnetic phase.

1. Introduction

Recently, rare earth based permanent-magnet materials have attracted broad attention due to their enormous applications in the fields of information technology, computer science and others [1,2,3]. Sm_2Co_{17} is a kind of high coercivity material acquiring the properties of high magnetocrystalline anisotropy, high Curie temperature, and excellent corrosion resistance and so on[4]. Later, many researchers have already prepared SmCo films relying on the physical or chemical method. Sayama et al.[5] synthesiezed SmCo films with perpendicular magnetic anisotropy on Cu underlayer by means of magnetron sputtering, and effects of Cu underlayer thickness on the structural and magnetic properties of SmCo films were measured. Some others pointed out that the structural and magnetic properties of SmCo films[6,7] strongly rely on the synthesized parameters. Zhang et al. reported that Ta buffer layer can be used to decrease the Cu surface roughness ,which enhanced the magnetic properties of SmCo films[8]. In this paper, Cr and Cu-doped Sm_2Co_{17} films on Cr buffer layer and Cu seed layer were synthesized using pulsed laser deposition (PLD). Results indicate that the incorporation of small amount of Cu and Cr can contribute to the growth of SmCo films and moderate Cr, which make the size of grain becomes smaller also enhance the coercivity of SmCo films. We also illustrate the magnetic exchange bias effect on Cu and Cr doped SmCo films.

2. Experiment

The $(Sm_2Co_{17}Cu_x)$/Cu/Cr (x=1.5, 2.5nm) films and $(Sm_2Co_{17}CuCr_y)$/Cu/Cr(y=1.6, 3.2nm) were synthesized on the Si(001) substrate using PLD. The purity of Sm, Cu, Cr and Co mental targets are 99.99%, and the base pressure was 2.0×10-4Pa. The film layer structure was $Sm_2Co_{17}Cu_x$ $(CuCr_y)$/Cu(100nm)/Cr(3nm)/Si(001). A 3-nm-thick Cr layer and 100-nm-thick Cu layer was a buffer layer and Cu seed layer, respectively. All samples were annealed at 680°C for 45 min, and the base pressure was 10-4 Pa. The surface morphology and structure of the films were measured by scanning electron microscopy (SEM) and X-ray diffraction (XRD), respectively. The hysteresis loops and temperature-depended magnetic properties of the films were measured using the Vibrating Sample Magnetometer(VSM) and Physical Property Measurement System(PPMS).

3. Results and discussion

The XRD patterns for $Sm_2Co_{17}Cu(1.5nm)Cr_y/Cu(100nm)/Cr(3nm)$ films (A y=0; B y=1.6nm) indicate that the major structure is Sm_2Co_{17}. The SEM images of Sm_2Co_{17} $Cu(1.5nm)Cr_y/Cu(100nm)/Cr(3nm)$ films (A y=0; B y=1.6nm) show that the grain size of pure Sm_2Co_{17} films is about 30 nm, and that of the Sm_2Co_{17} films mixed Cr(1.6 nm) is about 20 nm. This results indicate that incorporation of Cr not only refines the crystal grain, but also reduces its density. [9,10].

The hysteresis loops of $Sm_2Co_{17}Cu_x/Cu(100nm)/Cr(3nm)$ films are shown in Fig. 1. The coercivity of Sm_2Co_{17} layer doped 2.5nm Cu(1636 Oe) is smaller than that of Sm_2Co_{17} layer doped 1.5 nm Cu(1731 Oe). Their SEM spectra are showed in Fig. 2. The surface of the sample doped 1.5 nm Cu SEM image is more smoother than that of the sample mixed 2.5 nm Cu, and their average grain size is about 50 nm and 200 nm, respectively. We suggested that the superabundant Cu was not helpful to the growth of the structure of the Sm_2Co_{17} films[11].

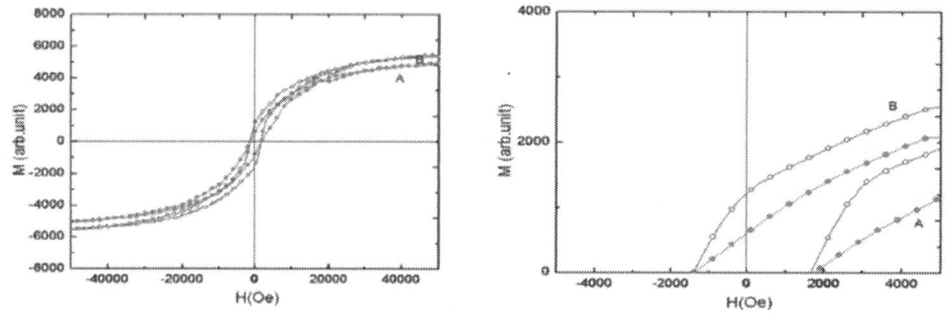

Fig. 1 Hysteresis loops of the $Sm_2Co_{17}Cu_x/Cu(100nm)/Cr(3nm)$ films (x=1.5nm A; 2.5nm B).

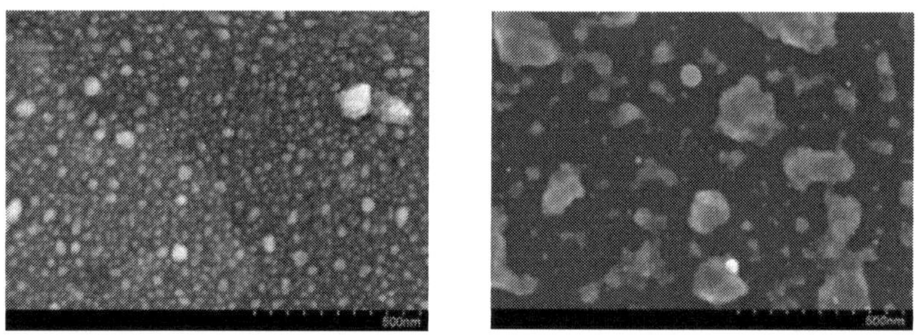

(A) (B)

Figure 2 SEM images of the $Sm_2Co_{17}Cu_x/Cu(100nm)/Cr(3nm)$ films (A, x=1.5nm; B, x=2.5nm)

Fig. 3 shows the hysteresis loops of the $Sm_2Co_{17}Cu(1.5nm)Cr_y/Cu(100nm)/Cr(3nm)$ films (A, y=0 ; B, y=1.6nm ; C , y=3.2nm). The coercivity of SmCo layer doped 0 nm, 1.6 nm and 3.2 nm Cr are 1731 Oe, 2120 Oe and 2554 Oe, respectively. Fig. 4 gives the curve of the coercivity properties related to Cr content. It can be seen that the coercivity of Cr doped films is higher than that of the films without Cr incorporation.

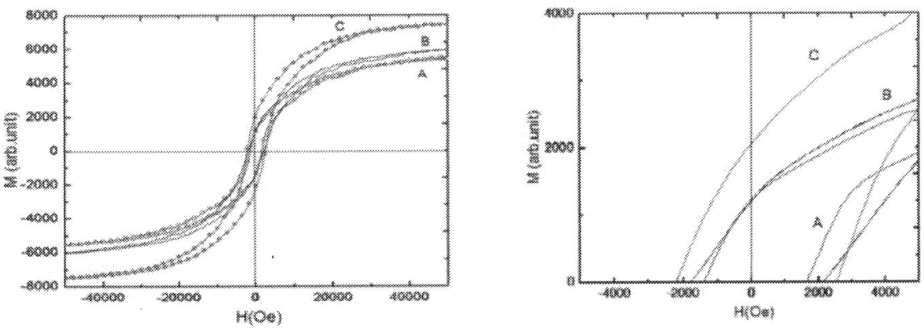

Fig. 3 Hysteresis loops of the $Sm_2Co_{17}Cu(1.5nm)Cr_y/Cu(100nm)/Cr(3nm)$ films (A, y=0 ; B, y=1.6nm ; C , y=3.2nm).

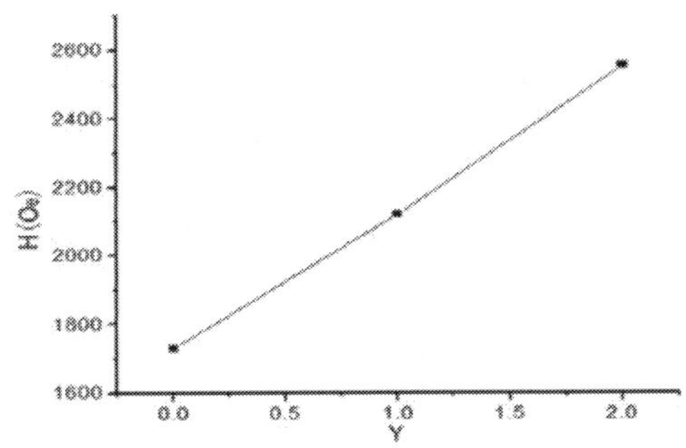

Figure 4 Dependence of the coercivity on Cr content for $Sm_2Co_{17}Cu(Cu1.5nm/Cr_y)$ /Cu(100nm)/Cr(3nm) films.

The SEM images of $Sm_2Co_{17}Cu(Cu\ 1.5nm)Cr_y/Cu(100nm)/Cr(3nm)$ (y=0 nm; 1.6 nm and 3.2 nm) are shown in Fig. 5. The grain sizes of the films doped Cr 0 nm, 1.6 nm and 3.2nm are about 10 nm, 16 nm, and 26 nm, respectively. This result indicates that doping Cr can contribute to reducing the surface roughness and grain sizes. The above studies suggested that opportune adjusting the content of Cr and Cu can optimize the magnetic properties and structure of Sm_2Co_{17} films[12,13,14].

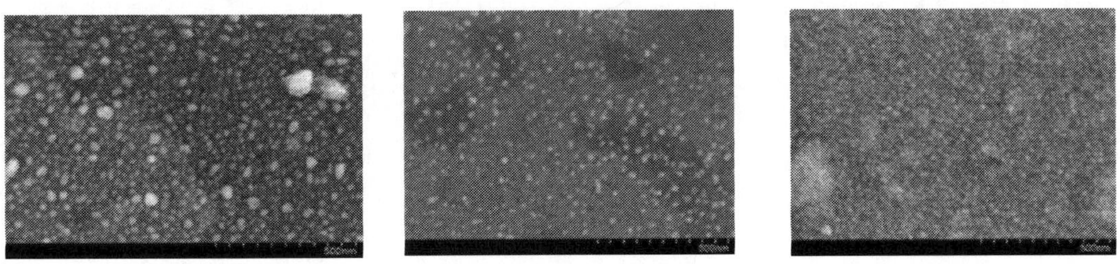

(A) (B) (C)

Fig. 5 SEM images of the $Sm_2Co_{17}Cu(1.5nm)Cr_y/Cu(100nm)/Cr(3nm)$ films (y=0 A; 1.6nm B; 3.2nm C).

We also measure the hysteresis loops of $Sm_2Co_{17}Cu_x/Cu(100\ nm)/Cr(3nm)$ (x=1.5 and 2.5nm) films, we find that the both hysteresis loops shift to right direction, indicating that there is a positive exchange bias effect in this films. Researches showed that the exchange bias effect mainly resulted from the antiferromagnetic-ferromagnetic interactions and the coupling between the magnetic moment and external field in magnetic interface. Exchange bias effect is usually negative in a state of cool field when the ferromagnetic- antiferromagnetic interactions demonstrate ferromagnetic

IOP Conf. Series: Journal of Physics: Conf. Series **827** (2017) 012018 doi:10.1088/1742-6596/827/1/012018

and vice versa. The appearance of positive exchange bias effect in Fig. 6 was attributed to the result of the competition between ferromagnetic phase (Sm_2Co_{17}) and anti-ferromagnetic phase (Cu) in the system [15,16,17].

4. Conclusion

In summary, $Sm_2Co_{17}Cu_x$/Cu/Cr(x=1.5nm,2.5nm) and Sm_2Co_{17}Cu(Cu=1.5nm)Cr_y/Cu(100nm) /Cr (3nm) (y=1.6nm,3.2nm) films are synthesized by pulsed laser deposition(PLD) method on the Si(100) substrate. The results indicated that doping moderate Cr could refine the grain as well as enhanced the coercivity, and that superfluous Cu would reduce the film density. Further study found that Cu and Cr dopant give rise to a positive exchange bias effect in Cu and Cr-doped Sm_2Co_{17} films. We suggested that the magnetic exchange bias effect was attributed to the interaction between antiferromagnetic state and ferromagnetic state in Cu and Cr-doped Sm2Co17 films and the intensity of exchange bias effect changed along with the competition between ferromagnetic phase (Sm_2Co_{17}) and antiferromagnetic phase (Cu,Cr) .

References

[1] Takei S, Shomura S, Morisako A, Matsumoto M, Haeiwa T 1997 *J. Appl. Phys.* **81** 4674.
[2] Liu J F, Chui T, Dimitrov D, Hadjipanayis G C 1998 *J. Appl. Phys.* **73** 3007.
[3] Neu V, Shaheen S A 1999 *J. Appl. Phys.* **86** 7006.
[4] Yan A, Bollero A, Müller K H, Gutfleisch O 2000 *J. Appl. Phys.* **91** 8825.
[5] Sayama J, Mizutani K 2006 *J. Magn. Magn. Mater.* **301** 271.
[6] Kato I, Takei S, Liu X X, and Morisako A 2006 *IEEE, Trans. Magn.*.**42** 2366.
[7] Sayama J, Miztutani K, Asahi T, Osaka T 2004 *Appl. Phys. Lett.* **85** 5640.
[8] Zhang L N, Chen J S, Ding J, et al. 2008 *J. Appl. Phys.* **103** 113908.
[9] Liu X H, Cui L Y, Zhou S X, et al. 2001 *Mater. Sci. Eng.* **304**, 1043.
[10] Andreescu R, O'Shea M J 2002 *J. Appl. Phys.* **91** 8183.
[11] Takei S, Morisako A, Matsumoto M 2004 *J. Magn. Magn. Mater.* **272** 1703.
[12] Ohtake M, Nukaga Y, Kirino F, *et al*, 2010 *J. Appl .Phys* **107** 09A706
[13] Zhao H B, Wang H, Liu X Q, *et al.* 2011 *J. Appl. Phys.***109** 07B715.
[14] Chen X, Hochstrat A, Borisov P, *et al.* 2006 *Appl. Phys. Lett.* **89** 202.
]15[Salaba E L, Rumplecker A, Kleitz F, *et al.* 2006 *Nano Lett.* **6** 2977.
[16] Nogues J, Schuller I K 1999 *J. Magn. Magn. Mater*, **192** 203.
[17] Zhou S M, Chien C L 2001 *Phys Rev B* **63**104.

Microstructure, magnetic and magneto-optical properties of Fe-doped $SmCo_5$ thin films

W W Wang, Q Q Fang, J J Yang, D S Dong and J G Li

Key Laboratory of Information Materials and Devices, School of Physics and Material Science, Anhui University, Hefei 230039, China

E-mail: physfangqq@126.com

Abstrat. The SmCo/Cu, SmCo/Cu/Cr and Fe-doped SmCo films on glass and Si (100) substrates are synthesized by pulsed laser deposition (PLD) technique. The microstructure, magnetism and magneto-optical Kerr effect (MOKE) of Fe-doped SmCo films are studied systematically. The results show that the corcivity and saturation magnetization of films deposited on Si substrates are better than that of on glass subatrates. It is also found that the surface morphology and magnetic properties of Fe:SmCo films are influenced by the annealing temperature. Meanwhile, the X-ray diffraction peaks of $SmCo_5$ phase are enhanced, especially for $SmCo_5$ (001), (002) and (003) peaks after annealing due to the enhancement of Cu (111) peak in the films with Cu buffer layer. Moreover, the MOKE of the samples on both glass and Si substrates are enhanced after Fe doping and the MOKE of samples is controlled by the Fe content, which provides a new way for optimizing the property of SmCo film as the magneto optical storages.

1. Introduction

Since the 1960's, SmCo alloys have been extensively studied. SmCo alloys have the high magnetic anisotropy properties (K_u), and the high Curie temperature (T_C), as well as the excellent corrosion resistance, high thermal stability and high magnetic energy product, which shows a good application prospect of SmCo alloys in high temperature environments and many fileds [1-6]. In 1989, Coehoorn *et al.* proposed the concept of nano-composite magnets based on the $Nd_2Fe_{14}B/Fe_3B$ phase [2], causing a wide attention of the academic community. In 1991, Kneller and Hawig proposed the concepts of exchange spring elasticity in the nano-composite model system [3]. They found that the magnetic energy product of nano-composite magnet can be greatly improved by the soft magnetic phase and hard magnetic phase in nano-composite materials. The reversible exchange interaction of the soft magnetic phase, Fe, is found in the low magnetic field and an irreversible exchange interaction of the hard magnetic phase, SmCo, occurs in the high magnetic field. The first-order reversal curve is usually used to distinguish the magnetization process of reversible and irreversible exchange interaction. For Fe:SmCo thin film, the current research focuses on how to enhance the magnetic properties and establish the exchange coupling model [4-7]. The study of Choi *et al.* shows that the interface effect of Fe:SmCo can be controlled by changing the thickness ratio of SmCo and Fe layers which increases anisotropy and maximum magnetic energy product [8,9]. MOKE technique is a fast and easy analysis way in micro-size arrange for magnet, recently, which has been widely used in the analysis of magnetic properties and MOKE [10,11].

We studied the characteristics of Fe:SmCo multilayer films with Cr and Cu as buffer layer, which prepared on the single crystal Si (100) and SiO_2 substrate by PLD. The magnetic properties and MOKE as a function of Fe concentration are studied under different substrates and annealing conditions.

2. Experiment

Fe-doped SmCo/Cu (Cr) films are prepared on glass and Si (100) substrates by PLD using a KrF excimer laser with a frequency of 8 Hz. The deposition chamber is pumped down to a base

pressure of 2.0×10^{-4} Pa and the substrate temperature is 400°C. The film thicknesses are controlled by deposition time. The film thickness is about 40 nm when the deposition time 40 min. The deposition time of Fe is set as 3, 5, 8 and 10 min, respectively, and corresponding samples deposited on Si or SiO_2 substrates are labeled as S1~S4 and G1~G4, respectively. The samples are annealed at the pressure of 5.0×10^{-5} Pa for 30 min. The annealing temperature of sample G4 and S4 are set as 650 °C and denoted as G4-650 and S4- 650, respectively.
The crystal structure is characterized by X-ray diffraction (XRD) with Cu K_α radiation. Surface morphology of the films is measured by scanning electron microscopy (SEM). MOKE curve is obtained using a high resolution magneto optical Kerr measurement system at room temperature.

3. Results and discussions

3.1. *Effect of Cu and Cr buffer layer on the SmCo thin film*

Figure 1 is the XRD spectra of SmCo (30 nm) /Cu (100 nm) and SmCo (100 nm) /Cr (30 nm) /Cu (3 nm) with different buffer layers of Cr and Cu on Si (100) substrates. In the range from 20° to 80°in Figure 1, the magnetic phase $SmCo_5$ (002) peak and Cu (111) peak are found. The stronger peaks of Cu (111) show that the Cu buffer layer has good texture, which is beneficial to the formation of the magnetic SmCo (001) layer because the high quality Cu (111) texture can promote the growth of $SmCo_5$ crystal structure. Sayama Junichi *et al.* [12] found that the coercive force of magnetic layer increased with the enhancement of Cu (111) texture and the studies of Ohtake and Nukaga *et al.* [13] have also given the same conclusion. In contrast to Figure 1(a) and 1(b), the diffraction peaks of the Cu buffer layer and the $SmCo_5$ layer are obviously enhanced when added 3 nm Cr layer on Cu buffer layer. Thus, the Cr buffer layer has promoted the growth of Cu (111) texture and the good crystallinity of $SmCo_5$. which agrees with Morisako's suggestions [14].

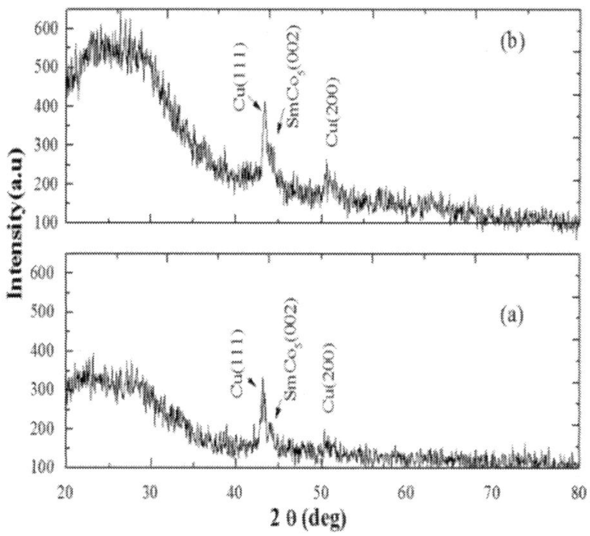

Figure 1. XRD measurements of (a) SmCo (30 nm)/Cu (100 nm); (b) SmCo (30 nm)/Cu (100 nm)/Cr (3 nm) thin film on Si (100) substrates

3.2. *Magnetic properties and MOKE*

The magneto-optic Kerr curves (or the MOKE magnetic hysteresis loop) of samples with different substrates as shown in Figure 2. For all samples, the MOKE curves dose not show "the wasp-waisted hysteresis loop", which indicates that the samples produced a magnetic exchange coupling with the Fe phase and SmCo phase. According to the exchange elastic model of Kneller and Hawig [3], a soft magnetic layer has a critical thickness

$$t_{cs} = 2\pi(A_s / 2K_u)^{1/2} \quad (1)$$

where A_s and K_u are a exchange energy of the soft magnetic phase and a magnetic anisotropy constant of the hard magnetic phase, respectively. If the soft magnetic layer thickness $t_s \leq t_{cs}$, the loop curve has a double phase behavior, while $t_s \geq t_{cs}$, the coupling of the soft magnetic layer and hard magnetic layer appears which result in a single hysteresis loop. For FeSmCo, A_s and K_h are 2.5×10^{-11} J/m and 1.71×10^{7} J/m^3, respectively, the critical thickness t_{cs} is 5.4 nm [15]. In our experiment, the maximum thickness of Fe layer is about 5 nm, obviously $t_s \leq t_{cs}$, hence, it satisfies the exchange coupling condition. We suggested that the exchange coupling of a soft magnetic phase Fe and a hard phase SmCo was formed in our samples.

Furthermore, the magnetic properties of the Si substrate samples are superior to the SiO_2 substrate as shown in Figure 2. For the eight samples, the coercivity values of SmCo films on Si (100) substrates are greater than that of on SiO_2 substrates. The coercivity values of SiO_2 (G1 ~ G4) are 48 Oe, 135 Oe, 241 Oe, 208 Oe, respectively, while that of Si (S1□S4) are 143 Oe, 176 Oe, 333 Oe, 351 Oe respectively. Obviously, the coercivity with Si substrate is greater than that of SiO_2 substrates for the same Fe content. In addition, for the same substrate the coercivity of samples increased with the increase of Fe content. Zhang *et al.* studied the effect of Fe on the coercivity of $SmCo_5$ as well as its formation mechanism to the $SmCo_5$ doped α-Fe with 0, 20, 30 and 35 wt% [7], which found that the coercivity can be controlled by domain wall pinning mechanism. Similar results are also shown in our measurements. We found that the increase of Fe content leads to the enhancement of the magnetic wall-pinning effect and promoted the increase of coercivity.

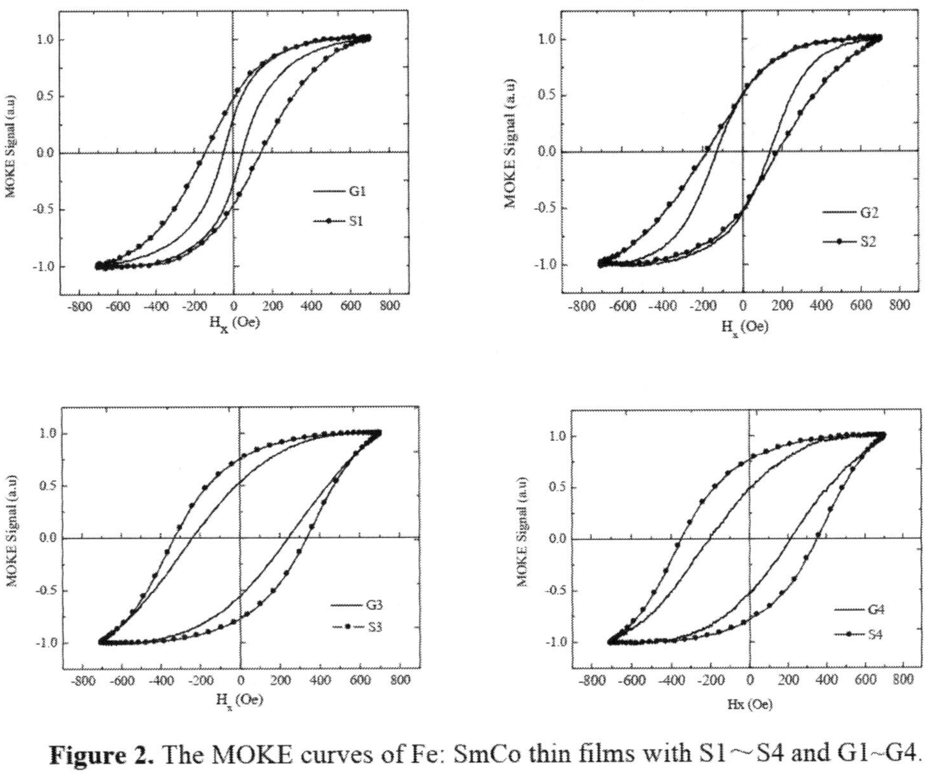

Figure 2. The MOKE curves of Fe: SmCo thin films with S1~S4 and G1~G4.

Figure 3 is the MOKE curves of the different samples deposited on Si (S1~S4) and SiO_2 (G1~G4) substrates. Fang *et al.* found that the Kerr rotation angle (θ_k), is proportional to the saturation magnetization [16]. The Kerr rotation angle is enhanced with the increase of Fe content, indicating that the addition of Fe is beneficial to enhance the Kerr effect in SmCo film. It means that the MOKE of samples can be controlled by the Fe content, which provides a new way for optimizing the performance of SmCo-based thin films as a magneto optical storage medium.

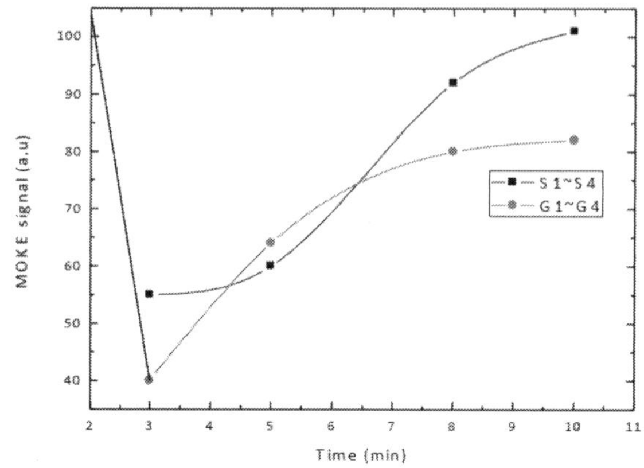

Figure 3. the change of Kerr effect with Fe content for the different substrate samples

3.3. Effect of annealing on the structure and magnetic properties

Figure 4 is the XRD spectra of the samples G4-650 and S4-650. The diffraction peaks of $SmCo_5$(002) and $SmCo_5$(200) appeared in sample G4-650, while $SmCo_5$(002) and $SmCo_5$(003) diffraction peaks appeared in sample S4-650. $SmCo_5$ (001), $SmCo_5$ (210), Sm_2Co_7 (008), Sm_2Co_7 (214) and Sm_2Co_{17} (304) diffraction peaks can be also seen (the highest peak corresponding to the Si substrate). Moreover, additional Cu (200) and Fe (111) diffraction peaks are seen comparing to the non annealed samples.

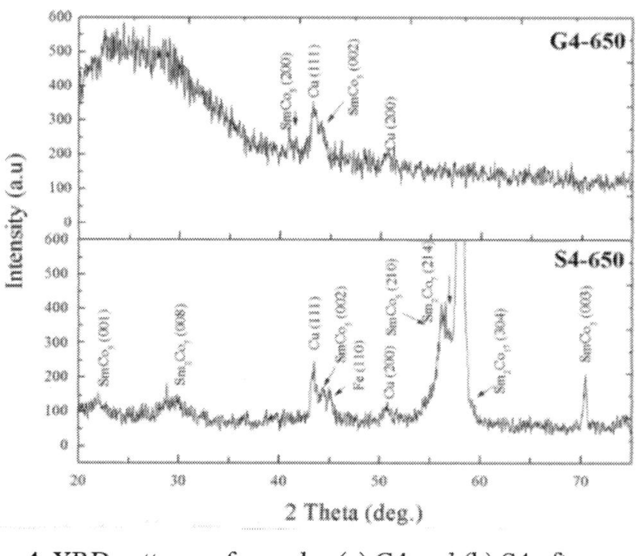

Figure 4. XRD patterns of samples (a) G4 and (b) S4 after annealing

Figure 5 gives the SEM pictures of the annealed sample G4-650 and S4-650. The average crystalline grain size of the sample G4-650 is less than that of sample S4-650. After annealing, the thin film surface is more uniform and without the large crystalline grain. It also shows that the sample surface is still relatively flat after 650 K annealing.

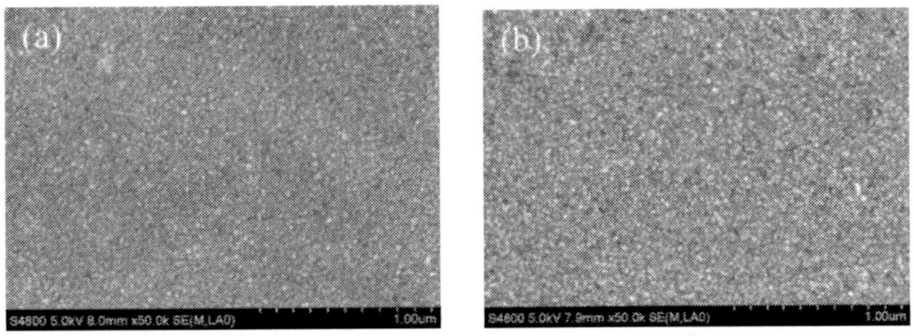

Figure 5. SEM spectra of samples (a) G4-650 and (b) S4-650.

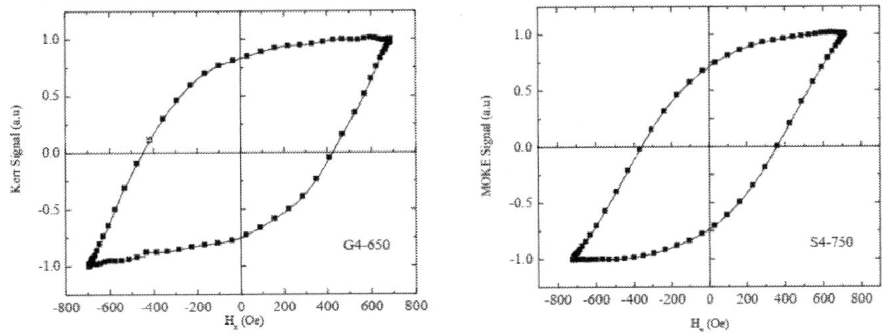

Figure 6. Magneto optic Kerr curve of sample after annealing (a) G4-650 and (b)S4-650

Figure 6 is the MOKE curves of sample G4-650 and S4-650. Contrast to Figure 2, the coercivity properties of G4-650 and S4-650 samples are significantly improved, and the coercivity values of S4 increase from 351 Oe to 370 Oe and from 208 Oe to 450 Oe after 650 K annealing, respectively. We suggested that the increase of coercivity after 650 K annealing is mainly due to the grain refine of samples, which is agreement with the reports of Zhang *et al.* [17].

4. conclusions

The magnetism, MOKE and the structure of Fe doped SmCo films were studied. It was found that the magnetism and MOKE of $SmCo_5$ phase increase with increasing Fe content. For the samples with different substrates, the magnetic properties of the samples on Si substrate are superior to that of on SiO_2 substrate, and after high temperature annealing, the increase of coercivity for both samples, of which the coercivity of SiO_2 substrate is better than the one of Si substrate. The experiment found that the crystalline grains of films were refine after the annealing, which makes the increase of coercivity. Moreover, the MOKE of the samples on glass and Si substrates are enhanced by the Fe content, which provides a new way for optimizing the MOKE of SmCo film as the magneto optical storages.

Acknowledgement
This work was supported by the Specialized Research Fund for the Doctoral Program of Higher Education of China and the Key Science Foundation of Education office in Anhui Province (GrantNo. 20120168).

IOP Conf. Series: Journal of Physics: Conf. Series **827** (2017) 012019 doi:10.1088/1742-6596/827/1/012019

Reference

[1]Sayama J, Mizutani K, Asahi T *et al.* 2004 *Appl. Phys. Lett.* **85** 5640
[2]Coehoorn R, de Mooij D B, de Waard C 1989 *J. Magn. Magn. Mater.* **80** 101
[3]Kneller E F and Hawig R 1991 *IEEE Trans. Magn.* **27** 3588
[4]Chu S Y, Majetich S A, Huang M Q *et al.* 2003 *J. Appl. Phys.* **93** 8146
[5]Fullerton E E, JiangJ SandBaderS D 1999 *J. Magn. Magn. Mater.* **200** 392
[6]Rong C B and Liu J P 2009 *J. Appl. Phys.* **105** 07A714
[7]Zhang J, Zhang S Y,Zhang H W *et al.* 2001 *J. Appl. Phys.* **89** 5601
[8]Suess D 2006 *Appl. Phys. Lett.* **89** 113105
[9]Choi Y, Jiang J S, Pearson J E *et al.* 2007 *Appl.Phys. Lett.* **91** 072509
[10]Wu K M, Wang J F, Chen K C *et al.* 2007 *J. Magn. Magn. Mater.* **310** e944
[11]Petracic O, Read D E and Cowburn R P 2007 *J. Appl. Phys.* **101** 09F510
[12]Sayama J, Mizutani K, Asahi T *et al.* 2006 *J. Magn. Magn. Mater.* **301** 271
[13]Ohtake M, Nukaga Y, et al. 2010 *J. Appl. Phys.***107** 09A706
[14]Morisako A, Kato I, Takei S, *et al.* 2006 *J. Magn. Magn. Mater.* **303** e274
[15]Zhang J, Takahashi Y K, Gopalan R, *et al.* 2005 *Appl. Phys. Lett.* **86** 122509
[16]Fang Q Q, Fang R Y,Zhang S, *et al.* 1999 *J. Appl. Phys.* **86** 3878
[17]Zhang L N, Hu J F, Chen J S *et al.* 2009 *J. Magn. Magn. Mater.* **321** 2643

Effect of high magnetic field on recrystallization behavior of cold rolled H70 brass

Y Wu[1], Q Wang[2] and X Zhao*[2]

[1] Research Institute, Northeastern University, Shenyang 110819, Liaoning Province, China

[2] Key Laboratory for Anisotropy and Texture of Materials (Ministry of Education), Northeastern University, Shenyang 110819, Liaoning Province, China

*Corresponding author: zhaox@mail.neu.edu.cn

Abstract. Effect of high magnetic field annealing on recrystallization behavior of cold rolled H70 brass was investigated. Specimens were annealed with and without a 12T magnetic field at different temperatures, respectively. The results showed that the recrystallization fraction of field-annealed specimen is lower than that of the non-field annealed specimen, while the percentage of ∑3 grain boundaries is higher than that of the non-field annealed specimen. The analysis results indicate that the high magnetic field hinders the recryatallization of cold rolled H70 brass, but promotes the formation of ∑3 grain boundaries. This may attributed to the formation of large number of asymmetric grain boundaries under the high magnetic field decreases the grain boundary migration process which leads to the slower of recrystallization process and makes the special grain boundary have more chance to grow up under the magnetic field.

1. Introduction

In recent years, the electromagnetic processing technology has been widely used in the solid phase transformation of materials. Researchers have paid many attentions to reveal the relationship between high magnetic field and the recrystallization behavior of the deformation alloys. Most researches in this field are related to magnetically affected recrystallization in ferromagnetic materials [1-3]. With the development of high magnetic field technologies, researchers found that non-ferromagnetic metals which almost have no reaction in the conventional magnetic field heat treatment are also quite sensitive under the high magnetic field. In 2003~2010, Molodov and his collaborators [4-9] explored the magnetically affected recrystallization and grain growth in non-ferromagnetic metals, and the results showed that the magnetic annealing can substantially affect the texture and grain structure evolution during recrystallization and grain growth in magnetically aluminum, zinc, titanium and zirconium. In 2008, Sheikh-Ali and his coworkers [10] found that the microhardness of diamagnetic pure copper, after the high magnetic field heat treatment, was higher than the specimen annealed without the magnetic field, they attribute it to the prevention of high magnetic field on the reduction of dislocation density in the specimens. In 2015, He et al [11-12] studied the effect of high magnetic field annealing on the recrystallization texture evolution in cold-rolled pure copper, and they found that the magnetic field annealing promotes the formation of the recrystalliation Cube texture in cold-rolled pure copper sheets. These researches have progressed the understanding of the high magnetic field mechanism on recrystallization in non-ferromagnetic metals. However, till now minimal work has been conducted to study the grain boundary character in non-ferromagnetic metals under the high magnetic field. In the present work, the recrystallization and grain boundary character of cold rolled H70 brass induced by the high magnetic field are discussed.

2. Materials and method

The material used in this study was a 50% cold-rolled commercially H70 brass plate, with a chemical composition of (wt%): 68.5.0~71.5 Cu, ≤0.03 Pb, ≤0.002 Sb, ≤0.5 Ni, ≤0.10 Fe, ≤0.002 Bi, ≤0.3 impurity, Bal. Zn. Specimens with the dimensions of 10mm×15mm×1mm were cut from the cold-rolled H70 brass plate, and were subjected to isothermal annealing at 300 °C, 400 °C, 500 °C and 600 °C for 30min, respectively, in a furnace installed in a cryo cooler-cooled superconducting magnet at a heating rate of 5°C /min, and then cooled in the furnace. During the magnetic annealing, a 12T magnetic field was applied during the whole heating and cooling processes. The specimens were placed at the center of the applied field with their rolling direction (RD) parallel to the magnetic field direction (MD).

Microstructures of longitudinal sections of the specimens were examined by orientation imaging microscopy (OIM). The scan was carried out over the area of 817×612 measuring points with a 0.25 μm step size. The percentage of recrystallized regions and ∑3 grain boundary were calculated using the Channel 5 software.

3. Results and discussion

Figure 1 shows the OIM maps of the specimens annealed at different temperatures with and without the high magnetic field. In these OIM maps, both the rolling direction and the magnetic field direction are horizontal. As figure 1 show, when annealed at 300°C, compared with the non-field annealed specimen, there still remains a small amount of deformation bands in the field annealed specimen. When the annealing temperature is higher than 400°C, the deformation bands completely disappear, and the grain size of the annealed specimens' increases obviously.

Figure 2 shows the recrystallization percentage and ∑3 grain boundary percentage of the specimens annealed at different temperatures, with and without the magnetic field. As figure 2(a) shows, the recrystallization percentage of field and non-field annealed specimens shows upward trend with the increase of annealing temperature. Compared with non-field annealed specimens, the recrystallization percentage of field annealed specimens are all lower than that of the non-field annealed specimens at the same annealing temperature. This indicates that high magnetic field annealing hinders the recrystallization of cold rolled H70 brass. As figure 2(b) shows, the percentage of ∑ 3 grain boundary of annealed specimens raises rapidly with the increase of annealing temperature. Compared with the non-field annealed specimens, the ∑ 3 grain boundary percentage of field annealed specimens is higher than the corresponding non-field annealed specimens at the same annealing temperature. This indicates that the high magnetic field promotes the formation of ∑3 grain boundaries.

During the magnetic field annealing, the diamagnetic brass interacts with the external magnetic field. Under the action of high magnetic field, the electron orbital of diamagnetic brass will change, and will produce a small magnetic field opposite to the direction of the external magnetic field, which will resist the action of external magnetic field.

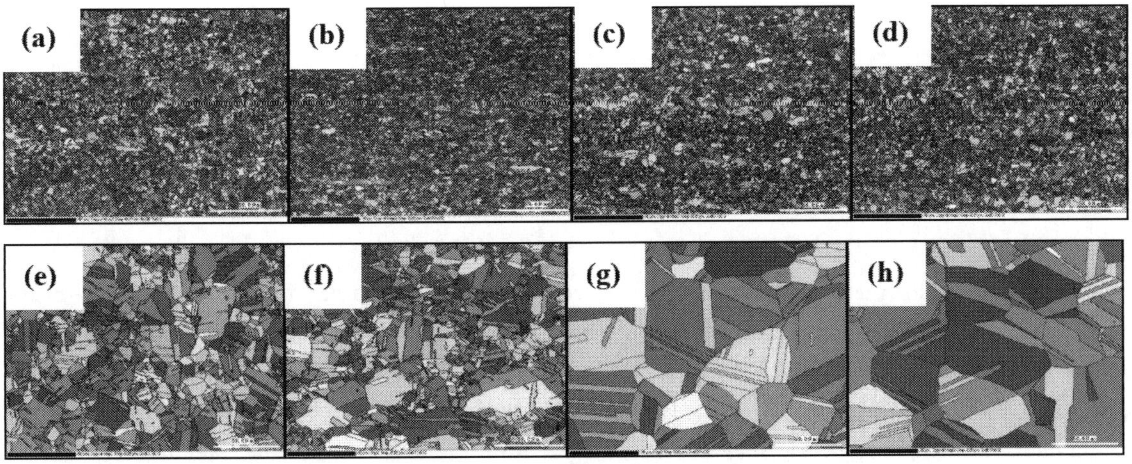

Figure 1. OIM（IPF+ Grain boundary）maps of specimens annealed at different temperatures.
(a) 300°C, B=0T; (b) 300°C, B=12T; (c) 400°C, B=0T; (d) 400°C, B=12T;
(e) 500°C, B=0T; (f) 500°C, B=12T; (g) 600°C, B=0T; (h) 600°C, B=12T

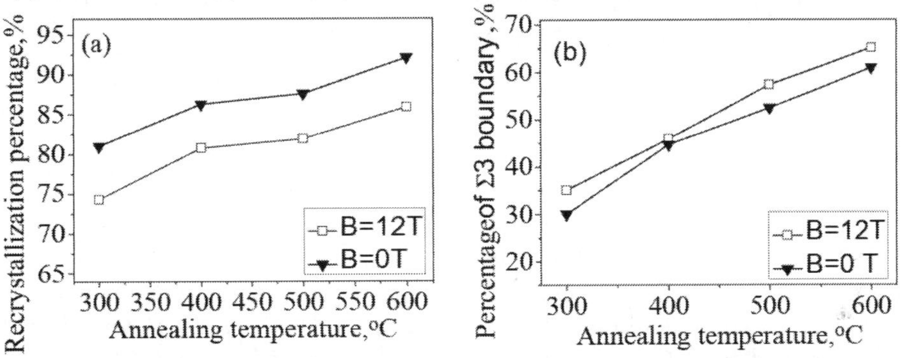

Figure 2. The percentage of recrystallized regions (a) and $\sum 3$ grain boundaries (b) in specimens annealed at different temperatures.

During the process of recrystallization, grain growth depends on the grain boundary migration process. For the recrystallization process under the high magnetic field, the movement of the grain boundary is influenced simultaneously by dual role of interfacial tension and magnetic force. Interfacial tensions at the grain boundary respectively point to the center of adjacent two grains, and their directions are opposite. While the magnetic force act at the grain boundary is related to the direction of the external magnetic field, its direction is fixed. So these two forces in different directions acted at the same grain boundary will make the boundary bend and redirect, thus the originally symmetrical grain boundary plane will become asymmetric tilt under the action of high magnetic field. Researches of Molodov [6] showed that the movement speed of bending tilt asymmetric grain boundaries during the annealing process is slower than the symmetric boundary under the same conditions. So, for field annealed specimens, the formation of a number of tilt asymmetric boundaries makes boundary migration become slow, so that recrystallization process becomes slow compared to the non-field annealed specimens. At the same time, $\sum 3$ CSL grain boundary has lower boundary energy compared with the random boundaries, under the high magnetic field, the grain boundary migration becomes slow, which make the special grain boundary groups have more chance to grow up. However, under the condition of non-field annealing, most of the grain boundaries (mainly high-energy large angle grain boundaries and slightly high-energy special boundaries) are in the metastable state. A large number of grain boundaries migrate at the same time, the growth of low-energy special grain boundaries is limited. Therefore, compared with the non-field annealed specimens, the percentage of ∑3 grain boundary in field annealed specimens are higher.

4. Conclusion

The effect of high magnetic field annealing on recrystallization behavior of cold rolled H70 brass process was studied. The results showed that the recrystallization fraction of field-annealed specimen is lower than that of the non-field annealed specimen, while the percentage of ∑3 grain boundaries is higher than the non-field annealed specimen. It indicates that high magnetic field annealing hinders the recrystallization process of H70 brass, but promotes the formation of ∑ 3 grain boundary.

5. Acknowledgement

The authors would like to express appreciation to the High Magnetic Field Laboratory of Northeastern University for providing the facilities. This work is supported by the National Natural Science Foundations of China (Grant No. 51271052) and the "111" Project (Grant No. B07015).

References

[1] Martikainen H O and Lindroos V K 1981 *Scand. J. Metall.* **10** 3–8
[2] Watanabe T, Suzuki Y, Tanii S and Oikawa H 1990 *Philos. Mag. Lett.* **62(1)** 9–17
[3] Enomoto M 2005 *Mater. Trans.* **46(6)** 1088–92
[4] Sheikh-Ali A D, Molodov D A and Garmestani H 2003 *Scripta Mater.* **48(5)** 483–8
[5] Molodov D A, Bhaumik S, Molodova X and Gottstein G 2006 *Scripta Mater.* **54** 2161–4
[6] Molodov D A, Bollmann Chr and Gottstein G 2007 *Mater. Sci. Eng. A* **467** 71–7
[7] Molodov D A and Bozzolo N 2010 *Acta Mater.* **58** 3568–81
[8] Bhaumik S, Molodova X, Molodov D A and Gottstein G 2006 *Scripta Mater.* **55** 995-8
[9] Molodov D A 2007 *Mater. Sci. Forum* **558-559** 649-56
[10] Sheikh-Ali A D and Garmestani H 2008 *Ceram. Trans.* **200** 413–9
[11] He T, Kang T Y, Wang Z J, Zhao X and Zuo L 2011 Mater. Sci. Forum **702-703** 411-14
[12] He T, Wang Y, Sun W and Zhao X 2015 *Mater. Sci. Eng.* **82** 012055 (1–5)

Institute of Physics
Dirac House, Temple Back
Bristol BS1 6BE UK

ISSN: 1742-6588
ISBN 978-1-5108-4075-1
9 781510 840751

CURRAN ASSOCIATES INC.
proceedings
.com